GIANT IN CHAINS

Books by Barrows Dunham

MAN AGAINST MYTH

GIANT IN CHAINS

Giant in Chains

by

BARROWS DUNHAM

Then the giant said, breaking his last chains, "Did ye think to hold me thus Forever: Ye fettered me while I slumbered, but none shall bind me now I am awake."

 HILL AND WANG · NEW YORK

Copyright 1953, by Barrows Dunham
All rights reserved
Library of Congress catalog card number: 52-12633

FIRST PAPERBACK EDITION SEPTEMBER 1965

Manufactured in the United States of America

TO MY WIFE, ALICE

When she is merry, then am I glad;
When she is sorry, then am I sad;
And cause why,
For he liveth not that loveth her
As well as I.

    A. GODWHEN: 15TH CENTURY

PREFACE

THIS BOOK, in some measure, shows a victory of logic over common sense. My purpose was to exhibit the fact that *all* the main problems of philosophy are involved in the simplest acts of personal and social life. It followed that I had to discuss the whole field.

To common sense it is obvious that not even a professional philosopher can perform this task with equal justice to every part. Nevertheless, I let logic run its course. During four years of labor, I carved away at the monolith, and reduced it at last to a shape and size which were at any rate discernible. Whether the shape is accurate and the size proper the reader himself must judge.

One of the troubles of composition, and one of its glories too, is the fact that when you write, you have to make up your mind. A publication of universal doubt would display not so much the complexity of the world as the perplexity of the writer. Within my theme, however, there were several subjects on which I had no decided opinion, and these I surrendered to the remorseless necessities of inference. I may hope that the position, if not persuasive, is at least clear. But clarity is an ideal scarcely less formidable than any other.

I gratefully acknowledge many generous permissions

to quote from books now under copyright. Specifically, my thanks go to Princeton University Press for passages from *A Kierkegaard Anthology,* and to the American–Scandinavian Foundation, which published Kierkegaard's *Concluding Unscientific Postscript,* where these passages originally appeared; to University of Chicago Press for a passage from *The Nature of Life* by Alfred North Whitehead; to Professor Clifford Barrett of Scripps College for passages from *Contemporary Idealism in America;* to Longmans, Green and Co., Inc., for passages from *Pragmatism, Essays in Radical Empiricism, A Pluralistic Universe,* and *The Will to Believe,* all by William James; to Harcourt, Brace and Company, Inc., for a passage from *The Acquisitive Society* by R. H. Tawney; to University of California Press for a passage from *The Rise of Scientific Philosophy* by Hans Reichenbach; to Oxford University Press, Inc., for a passage from *Appearance and Reality* by F. H. Bradley; to the representatives of the late Bernard Bosanquet for passages from *The Value and Destiny of the Individual, The Philosophical Theory of the State,* and *Social and International Ideals,* all by Bernard Bosanquet; to Macmillan and Company, Ltd., London, for passages from Bosanquet's *The Philosophical Theory of the State;* and to the Macmillan Company, New York, for a passage from *A History of Modern Philosophy* by William Kelley Wright.

The habit of such acknowledgments, which a long practice does render somewhat formal, serves nevertheless to show how many people it takes to make a philosophy. I think the fact is fortunate, for to my mind the most endearing thing about philosophy is its source in people. What I see there is not alone the skillful logic of genius or (as sometimes happens) the prophetic pulse of poetry, but the dim uncertainties, the yearnings, the struggles (and even the inertia) of multitudes. Of all these the famous philosophies are but refinements; and the famous philosophers, however lofty in view or detached in temperament, convince me of nothing so much as of the fact that they are men.

BARROWS DUNHAM

Cynwyd, Pennsylvania
December 10, 1952

PREFACE TO THE PAPERBACK EDITION

This book, of the year 1953, was a sort of sequel to *Man Against Myth*, of 1947. In *Man Against Myth* I had subjected certain familiar notions to philosophical criticism: "You can't change human nature" and "Thinking makes it so" were two of these. The book had some success—chiefly, I think, because people were surprised to find that such notions really would collapse under analysis. I thought it fair to conclude that I had shown the power of philosophy to rid mankind of cramping or defeatist ideas.

An old maxim in philosophy asserts, very truly, that to deny is to affirm. If philosophy can thus dispel illusion, it can also tell us, and must try to tell us, the great realities. These are, in general, three: what the world is, how we know the world, and how we ought to act in the world. I thought, therefore, that the author of *Man Against Myth* might reasonably proceed to the larger, more difficult questions and hope to take his readers with him.

The attempt was not easy. For some sixty years, professional philosophers have attended to only one of the great realities—namely, "how we know." They have surrendered "what the world is" to the scientists, and "how we ought to act" to the clergy and the novelists as a muddled and possibly meaningless question. Such shrinkage was quite recent: from the fifth century B.C. to the nineteenth century A.D., philosophers had discoursed upon the three themes as inseparable.

Taken together, these themes are the single theme of man and his place in nature. I have all my life thought, and I still think, that *this* is the subject of philosophy. But it must be admitted (for it is the fact) that there are not many philosophers in the West who think so. Accordingly, I felt a little lonely in making the effort; and this feeling grew upon the further fact that the Cold War, which began about the same time, professed an ideology I would have to oppose. I was a mere man facing a mighty myth.

It chanced that, during the presidential campaign of 1948, the Progressive Party had a broadcast in which it presented the recorded words of Mrs. Isaiah Nixon, a Negro citizen of Georgia. Her husband had been shot and killed for voting in the primary election. She spoke a little haltingly, but every pause declared the extraordinary eloquence of the whole. And there was the language, which rose, as language should do, from a knowledge of the world as it is, joined with a knowledge of what the world should be. It was the pure Shakespearean tongue. When I heard it, and heard her, I had my book.

Thus, I have presented philosophy not as *an* but as *the* answer to Mrs. Isaiah Nixon's questions—as the theory of human deliverance, as the ultimate resource for human control over society and the world. The

PREFACE TO THE PAPERBACK EDITION xiii

Giant of my title is, of course, mankind; and philosophy is at least part of the means by which the Giant can break his chains. I may add that the apparent quotation on the title page is a sentence of my own composing.

The climate of opinion into which this book issued was by no means of the sort an author hopes for. By 1953, most thinkers had vanished into a mist of safe opinions; and the clamor of those years, loud though it was, gave no more description of reality than silence would have done. Yet, along with the troubles, there were joys. Two anecdotes will suffice for illustration.

I had originally placed in Chapter III some five or six quotations from Josiah Royce, so as to give, in addition to the content, the peculiar flavor of his style. The quotations were so brief that there was no legal necessity for permission to use them. Nevertheless, I wrote to the owner of the copyright and asked for permission to quote. The owner, who had once been, I believe, the baby mentioned on page 50, replied that before he would grant permission, he would have to know whether I was a "collectivist." Franklin Roosevelt, he said, had described Josiah Royce as an inspirer of the New Deal; and ever since then, he, as a loyal son, had sought to lift that taint from his father's name. For my part, I was not prepared to submit to any such inquisition. Accordingly, I withdrew my request, and turned all the quotations into indirect discourse. Thus the book lost its little bursts of Roycean prose, a prose that remains one of the curiosities of literature.

There are, however, other sorts of men. In 1949, I wrote a fan letter to Sean O'Casey, after reading the fourth (and best) volume of his autobiography. He replied with some kind words about *Man Against Myth,* which had appeared in England. Sean had a taste remarkable among Europeans: he liked Americans. Moreover, he was extremely generous about helping younger writers. Sometime in 1952, I reported to him, hyperbolically, that he was delaying the composition of *Giant in Chains.* I had been charmed by his trick of making statues talk (for example, the statue of Saint Patrick has a long and very funny debate with Bernard Shaw), and I was hoping to hit upon some similar device. His reply was, "Don't let me stand in your way."

Every now and then, if one is lucky, something happens that makes everything plain. I think that a complete theory of ethics, and perhaps a theory of knowledge too, can be got from that simple, gracious sentence, "Don't let me stand in your way." *Giant in Chains* attempts both these theories, but I am not so senseless as to suppose that I have succeeded where all previous philosophers have failed. The way, however, does seem open, and it has the added charm that readers may traverse it as far and as freely as they wish.

BARROWS DUNHAM

Cynwyd, Pennsylvania
July, 1965

CONTENTS

Preface vii
Preface to Paperback Edition xi

Introduction

I   Then Why Not Every Man?   3
II   Philosophy and Mrs. Nixon   19

This Part of the Country

III   Permanence without Change   45
IV   Change without Permanence   81
V   Permanence through Change   102

Know Just Why

VI   The Know-Nothings and the Know-It-Alls   143
VII   Queen Truth and King Charles   170
VIII   There Is Method In't   189

Such a Hard Time

IX   The Candle and the Sun   209
X   The Higher Height   236

Index   263

INTRODUCTION

CHAPTER I **THEN WHY NOT EVERY MAN?**

WHEN THE SUN hangs lower than the lowest branches, the world, much wearied, slides out of its day-long rut, and people move by various paths to their release. It has been a hard day, stirred with the little excitements of failure or success: eight hours of one's life, spent (who knows how fruitfully?) in acquiring the means of working another eight.

The man in the street, who is often quoted and has never said a word, is at last — in the street.

He walks past the bank that has his money and keeps it safe from him by closing at three, past the shop windows whose reductions he can't afford, past the bootblack-hatcleaner who beautifies both extremities, past the bars where desperate men are drinking and the literary assassins lurk. He takes a bus or train or trolley, and, after a forgetful interval, he comes home.

And here, if not quite heaven, there is a haven: the loved, enduring sameness which all the motions of his life go to sustain, the immediate hearth of values, the rock and tower whence he looks out, protected, upon the world. He is always building it, in fancy and in fact. Looking backward, he can see the history of half-unnoticed change through which it has survived and even grown. Looking forward, he can hopefully guess the ampler wage, the more rewarded talent, which is to make all happiness secure. And beyond that, his own ultimate, the very alteration of his being — not storm or terror, he will wish, but a simple leaving off of light.

Doubtless he does not always range through all these notions. Yet in each opening of the door there are the inevitable questions: Are they at home? And well? And happy? If so, then life may wander as it will until another evening and the questions come again.

The man in the street, now the man at home — who is nearer than he to the subtle conflict of permanence and change? Who knows, more than he, what effort it takes to make possession last, to hold within the moving universe some firm abode? Surely he is shrewder than Parmenides, who thought that logic had abolished change; sturdier than Hume, who found in all the fabric of the world mere causeless series of sensations.

And now at home, he shuts the door upon the outer world but not upon philosophy. Seated at table and exercising his right to the day's secrets, he learns everything that happened, what it was and how it felt. Though the notion may not rise into full consciousness, his behavior suggests an understanding that events such as these are what the world is really made of — not the gossip of headlines nor the loud obscurities of commentators, but the effort of people like his wife and children to do what they think ought to be done at the moment they have to do it.

Now, this is a generalization to the effect that history is made by people. It is not particularly a favorite with historians or philosophers, who have their eyes (let us say) upon the splendors and lusts of kings. Nevertheless, it is simple, it is very arguable, and it is in all probability true. The man at home, if modesty did not so shroud him, might marvel how many learned errors he has escaped.

But his acumen is not yet exhausted. For I fancy that as he reads his evening paper, backwards, from the comics through murder, theft, fire, and divorce to the serried horrors of Page One, he rejects this item, accepts that, rather doubts the other, and in general wonders how much correspondence there may be between the story and the event. He began life, to be sure, believing what he was told; but he has learned that many things which are told are erroneous, and that one must do a little testing for oneself. One seldom, perhaps, fully develops the technique of inquiry. One does know, however, that contradictions within a story will destroy it, and that prejudice will tear it loose from the sustaining facts.

These are the rudiments, indeed the essentials, of a theory of knowledge — a theory, that is to say, which should be able to give us the talent of true belief. We all have some impression of what this theory contains, for the total lack of it would leave us helpless; and very probably we all feel that, however much we now know of it, there would be benefit in knowing more.

The man at home, soon to be the man in bed, has thus spent an entire evening without being released from philosophy. His last reflections are set by an item in the paper to the effect that a lecturer at the Good Feeling Club has exhorted the audience not to care for material well-being, which is equally attainable by rabbits and snakes. But, on this supposition, one has labored for inferior substance, and all the iterated tasks were frauds. Moreover, shall one conclude that, in this fragment of the universe, what one buys with so much toil has really lesser value, while pearls abound for simple taking? That would be a paradox indeed, to make faint sense of. And thus over the long day ethics slides her curtain, the first and deepening purple of the night.

Philosophy As Reflection

Unless I have misconceived the habits of homes and of evenings, I am entitled to my conclusion that it is very natural for men to philosophize and that they do it oftener than they know. They do it, moreover, better than they know, just because they do it naturally. They are not asked to publish their findings, and hence are not tempted to the platitude and obscurity of official pronouncements. Nor do they dispense that purchasable comfort which sighs and lies and lets the misery last.

On the contrary, they are simply trying to make their lives intelligible, a process they began in their cradles. If they are amateurs rather than professionals, they are so in a profound sense, namely, that they love understanding. They practice philosophy in its ancient and traditional form, as reflection.

And what is it to reflect? It is to draw the scattered data of experience into various unities, to find the likenesses and contrasts, to catch the logic moving throughout change. A sunrise and a sunset make one day, and of such days a handful makes a life. Yet in these regularities, dull (it has sometimes seemed) as the ticking of a clock, lies the vast stretch of objects to be known.

The simplest facts are pregnant with philosophy. Men, it is obvious enough, are born of their own kind, are reared among men, and die among men. The relations thus instituted are somewhat shifting, somewhat insecure, and at their extremes rapturous or tragic. During a lifetime, every man stores up some knowledge of them and some theories about them. Thus, despite prejudice and naïveté, every man is, at least in rudiments, a social scientist.

Secondly, men are everywhere and always in some relation with the physical world, which is the source of food, clothing, shelter, and the innumerable other commodities they need. During a lifetime, they acquire knowledge and develop theories about this world also. They are, in rather more than rudiments, physical scientists.

Now, it happens that between these worlds, the physical and the social, there are many interactions. The technology of producing goods is mainly a problem in physical science, and the distribution of those goods is mainly a problem in social science. Nevertheless, the two profoundly influence each other. Failures in the scheme of distribution will thwart technology and sometimes bring it to a halt. On the other hand, a failure (and sometimes a great success) in technology will unsettle all social relations. A crisis in either world begets a crisis in the other; and the human race, now master of the atom, stands terrified, knowing that it could be affluent and happy, knowing also that it may perish altogether.

Since in man these two worlds meet with cordiality or with violence, it is clear that there is a kind of knowledge incorporating what is known about society and about physical nature, but larger, more nearly universal than these. It will consist of statements describing the whole complex of relations in which we stand, the welter (as it may appear) of happenings which surge around us, jostling, pushing, driving us upon our destiny. It will consist, further, of statements asserting what goals we seek or ought to seek. And lastly there will be statements composing the master plan by which we are to direct, so far as we may, the whole great process to our human goal.

This congeries of statements — this *system* of them, as we must hope it may become — is philosophy. Its single theme is Man and His Place in Nature. Around this theme are gathered the clusters of knowledge and theory called, in the darkened language of tradition, ontology, epistemology, and ethics. Which words, taken all together, are a deafening way of saying that before mankind attains its ultimate safety, we shall need to know what the world is, how it is known, and to what ends it ought to be controlled.

Generalizations must be large to encompass such subjects. Their very size invites abstractness, as if it were their fate to obscure the labors of lesser but more lively men. This notion is due in part to the fact that the first acknowledged philosophers were leisured aristocrats and the latest are university professors.

It is also due, I fancy, to a certain aloof exercise of

philosophical techniques. The thinker, demonstrating his claim to be the "spectator of all time and all existence," abstracts some attribute from the mass and hangs it like a blanket across the stars. Or, preferring the little world to the great, he may show us infinity in the palm of his hand, where the touch is thrilling but the shape is odd.

Though I seem reproachful of these exercises, the truth is that I know them and love them well. They are the sinews of meditation and are no less active for being calm. Scientists, to be sure, have public motion, even hurly-burly, for they are to be found in laboratories, looking through microscopes and into test tubes, and handling with a fine boldness much combustible material. Philosophers are, by habit and by reputation, a much quieter breed. They will be found in armchairs, and, when they are not found in armchairs, they will be found in bed. But this only means that some generalizations can be formulated seated or prone, and perhaps must be formulated that way.

Such, we say, are the habits of philosophers; but I think there are familiar moments when the process shows itself in every life. The mood, which now and then descends, of "What is it all about? What am I doing here?" differs only in precision from Kant's celebrated questions, "What can I know? What ought I to do? What may I hope?"

It is the genuinely philosophic mood, into which, as most people have it, float scraps of ill-digested sermons, pedagogic homilies, and calendar mottoes. But the authentic content is various insights which experience has suggested and a native intelligence has made shrewd.

It is this that gives such pith and pregnancy to folk sayings and emboldens us to hope that the world, which plainly will not be saved by its scientists, will be saved by its people.

Philosophy As Guide

"Saved by its people" — the phrase suggests that philosophy is more than reflection. For if its people are to save the world by means of some wisdom original with them, they will have to philosophize with a view to guiding their actions. Thus philosophy gives eyes to practice, and practice informs philosophy.

That philosophy is the guide of life is an old boast, not limited to philosophers. It is sententious enough to invite agreement, and it can be made by men whose lives are not conspicuously guided in this way. Yet a precept coined by the richest of Athenian intellects and sanctified by the sufferings of Spinoza will not be much tarnished by inferior use.

There is reason to think, for example, that the philosopher-kings, rather too aristocratically defined by Plato, are an ultimate historical necessity. They will be, not kings, but a commonwealth of knowledgeable persons accustomed to settling problems by general principles. They will appear in that not unimaginable epoch when human science, no longer spent upon fattening an elite, brings treasures to every door.

Meanwhile philosophy needs some effort to make its value known. Part of that value, I suppose, is already recognized in Horace's maxim,

> Aequam memento rebus in arduis
> Servare mentem.

But calm in perplexity and courage under loss are attributes mainly personal. Now that everyone knows how closely each life is linked with others, how impossible the chance of living to oneself, philosophy appears anew, as guide to social action.

For such a use its very substance fits it. That substance, which we have called "Man and His Place in Nature," is also the stuff of politics, for man's political behavior is simply his effort to determine, by conflict as by concord, what his place in nature shall be. To this great theme no science or combination of sciences is adequate, for their generalizations are too small. No art or combination of arts suffices, for their skills govern only segments of the whole. The talent for discussing (so far as may be) all things and all relations, the given and the desired, the means and the end, is philosophy's alone.

If philosophy could not claim leadership by right of content, it would nevertheless acquire it by surrender. The sciences have for many years been as explicit concerning what their content is *not* as concerning what it is. They say, for example, and quite falsely, that they have nothing to do with ethics. The natural sciences say that they have nothing to do with politics — an illusion which not even the hydrogen bomb seems able to explode. And the social sciences, in terror of harboring dangerous knowledge, are beginning to talk as if they had no knowledge at all.

History, however, will not leave undone what scientists are thus neglecting to do. Despite the undoubted possibility, it seems improbable that mankind will permit science to work universal destruction. In conquering at last these lethal uses, mankind will work out the theory and practice which can adapt science entirely to human welfare. The theory and practice thus attained will mark the passage from philosophy militant to philosophy triumphant. I do not suggest that professional philosophers as we now know them are likely to achieve all this, but I do say that it will be achieved and that the men who achieve it may be called philosophers.

Philosophy As Conflict

Politics rules everything except economics, and economics rules politics. Beneath these lordly hierarchs all other disciplines must find their place. Throughout past ages, philosophy has been partly what philosophers wanted it to be, but mostly what rulers permitted it to be. Its recurrent mood of aloofness from politics was always illusory and masked some deeper political fact. Such illusions cannot be trifling. Men who are self-deceived are not likely to undeceive others, and their theories will be but treacherous guides.

Consequently, the professional and the amateur and the man who treats these problems unawares must learn what may seem a dreadful fact: it is in the fires of politics that philosophy is to be forged. It will be shaped in conflict and brought forth in whatever victory the conflict may yield. Thus molded and produced, it may or may not contain various true statements; but it will be the philosophy which has prevailed, and it may seem to be the only philosophy there is.

The fires I speak of are bright and hot, and will suffice to melt the weaker metal. For politics has this peculiarity: although in function it is the noblest of human activities, it is sometimes in use the basest of all. It is noblest in function, because it alone can bring about the general welfare of mankind. It is basest in use, when it affords the place and means by which men whose interests are not public interests can compel obedience to their rule.

The politics through which our philosophy must pass is conducted by skillful veterans, who regard morality, not as distasteful, but as profoundly inconvenient. Despite their threats of universal slaughter, they are not misanthropes: that is too absolute a view for politics. They are, rather, the representatives of a class whose hold upon world events is slipping and might be permanently dislodged by controversy at home.

That class, the capitalist, having monopolies at home and investments abroad, feels upon these the erosion of history. Nothing can be more natural than that it should wish to consolidate its power. Although it puts more faith in machines than in people, it is by no means indifferent to what people think. And since what people ultimately think is philosophical, not a little of the general pressure for conformity is felt in philosophy itself. Consequently, the capitalists give daily and empirically a forceful and even violent demonstration of Marx's old thesis that "the ruling ideas of each age have ever been the ideas of its ruling class."

Such circumstances invite a frank use of all the instruments of coercion. Blackmail, slander, loss of livelihood, imprisonment, and penalties perhaps still worse are to be the means by which conformity prevails. Dissent grows costlier to the dissenter, who must harden himself for the agonies of public execration and the torments of physical abuse. The range of "free" (that is to say, safe) thought narrows to a few statements, and on this inconsiderable theme the timid and the mercenary write tedious variations.

At first sight, no condition seems less suited to philosophy. Truth, a primary aim, is caught up in the strife of factions. Every lure to passion and mendacity exerts its power, and rival thinkers seem less engaged in understanding the world than in abusing one another. "Truly," said Spinoza, rebuking the contentious clerics of his day, "if they had one spark of light divine, they would not rave so arrogantly, but would learn to cherish God with greater wisdom, and be as remarkable among their fellow men for love as they now are for malice." [1]

Nevertheless, Spinoza's own century shows that violent social struggles are quite compatible with philosophic advance. An epoch which can boast the names of Descartes and Leibniz, Hobbes and Locke, Newton and Boyle, was surely not a time of universal mist. There is, in fact, much reason to suppose that the very sharpness of the struggles refined the philosophy.

And the reason is not difficult to grasp. For one thing, the same motive which excites distortion also excites a closer attention to what is real. In every such crisis the contending parties need to justify themselves, but they need, above all, to know exactly where they stand. The risks of self-deception are enormous. Consequently, the warping of facts for the sake of propaganda is much

[1] *Tractatus Theologico-Politicus,* Preface. (My translation)

modified by the necessity of knowing what the facts are.

Now, "knowing what the facts are" involves more than the mere compiling of data, fact by fact. There must also be authentic interpretation of the data, so that the pattern of events is clear. This in turn raises questions of logic and method, which must be solved or at least advanced toward solution. What began as a contest for political power ends by involving the whole range of philosophical discussion.

Moreover, it does so on a mass scale. Each party is engaged in mobilizing its forces, which, in the last analysis, are *people*. The unity on each side is in part sustained by the sharing of beliefs, that is to say, by a common ideology. These ideologies are often more inventive than true. Nevertheless, it is the case that the success of an ideology does, in the end, bear close relation to its scientific validity. The fascist propensity for nonsense ended in disaster, and all its later imitations will do the same.

These results are surely what we might expect. If we assume (as an abstract and ideal condition) a social movement having full knowledge what aims are attainable and what means are adequate, its victory would be a foregone conclusion. Any opposing movement would be less knowledgeable in leadership and membership, and therefore weaker.

The other extreme is, unfortunately, neither abstract nor ideal. For there are, as there have been, social movements such that if the members knew what the movement really is, they would desert it. Such movements, of course, are doomed. They exist upon illusion, only to be shattered upon fact. Despite all boasting, the Great Lie has never been a permanent historical success.

Knowledge is man's contact with reality and the ground of his control over it. Therefore, the more knowledge a social movement has and the more that knowledge is diffused among the members, the greater the probability of success. Those who can predict the future, and who want what they predict, are on the future's side and will prevail. Seen thus, every social crisis is a time of great discovery, in which the world, once blurred by inattention and deceit, breaks the ambiguous silence to affirm, "This is the real, and this the true; these are the goals to be desired."

In thus describing the effect of crisis upon philosophy, I intend no praise of crisis as such. It may well be that one philosophizes better in "calm of mind, all passion spent." Certainly one philosophizes then more pleasantly. But circumstances so Arcadian are not in these days to be found. We must take our epoch as it is, and, seeking in it the knowledge to resolve crisis, establish a connection between that knowledge and our hope.

Now, hope, as I conceive it, means that events can be made to issue in favor of the world's great multitudes, who are the chief victims of social violence. By knowledge I mean that theory which, having the rational attributes of a science, is able to describe the crisis and propound a solution. If our account has been correct, we can say that the theory of all such descriptions and solutions is philosophy, and that therefore philosophy is the one discipline which is concrete and comprehensive at the same time.

Our purpose in these pages, then, is to consider philosophy as the theory of human deliverance — of deliverance, that is to say, for mankind as a whole. The impoverished many are to be freed from economic and cultural bondage; their masters, the few, are to be freed from the guilty unease and degraded illusions to which their status condemns them. Surveying the course of social development, one can scarcely avoid thinking that such is the probable outcome of present events. The age of privilege is ending, that long, long age of some six thousand years; and in the centuries to follow, men will share the planning of their happiness, even as they will share the happiness they plan.

Philosophy is deliverance or it is nothing. It saves everybody or it saves none. It cannot be shrunk into parcels or laid away in libraries and desks, for, no matter what the professionals think philosophy to be, their actions and other people's actions are all the time determining what philosophy in fact is.

Consequently, our man in the street, whose bent toward philosophy we saw with some surprise, was simply what he had to be. He could not make the smallest move without engaging the largest questions. He could not know his own welfare unless he also knew something about knowing and about welfare. He could not act for either unless he knew something about change. A little awkwardness, and the goal escaped him; a little deftness, and the prize was won.

It has not been otherwise with the great philosophers — with Socrates, Plato, Aristotle, Augustine, Spinoza, Locke, or Kant. From low to high, speculation issues from a common need and meets a common success whenever theory and practice harmonize. The best thinkers have never hoarded wisdom. What they are willing to share we need not hesitate to receive.

"Didn't my Lord deliver Daniel?" asks the old spiritual. And it answers, "Then why not every man?"

CHAPTER II PHILOSOPHY AND MRS. NIXON

A MAN UPRIGHT OF LIFE, unsoiled by crime, need fear (Horace says) no hostile darts and may even have a Lalage to sing of. A thousand schoolmasters, headmasters, and masters of arts, whose leg the old poet perennially pulls, have intoned the splendor of that opening stanza and then slid by the others, blushing and unconscious of the jest. For, in their view, a serious and radical incompatibility exists between *integer vitae* and *dulce ridentem,* between the upright life and a woman's smile.

They are wrong, of course, and they would have proved sad corrupters of our youth, if nature herself had not spoken more accurately and imperiously. Once the book has closed upon its page and the room upon the lecture, we rush, so to say, out of doors to seek what may be found of joy and comradeship and love. Or at least we naturally do so unless a habit of gross self-interest has destroyed in us the taste for these delights.

We are set down, it appears, in a world where such goods are obtainable, though never without knowledge and seldom without toil. The ancient peoples, who had a good deal more toil than knowledge, were of the opinion that the physical universe maliciously concealed itself. "Nature loves to hide," said Heraclitus bitterly; and the Egyptians inscribed upon the temple of Isis, their nature goddess, the legend, "No mortal hath lifted my veil." There was a dread of knowing too much, and even a sense of guilt at discovering anything.

Yet despite these timidities, in which may be discerned the terror of rulers lest their conditions of rule be finally revealed, philosophy took a shape classical and never to be lost. From Thales to Anaxagoras, the pre-Socratic philosophers discussed the physical world — discussed it, indeed, materialistically. Thus far philosophy did not consciously exhibit the fact that the thinkers lived in society. But when Protagoras, on behalf of a self-assertive ethics, wished to undermine the whole system of custom and use, he attacked not moral knowledge merely but knowledge in general. Only those sentences could be true, he thought, which stated the experience of one particular man at one particular moment. No *universal* truths, therefore, no maxims, no rules, no principles.

The doctrine has been much applauded in our own day and for much the same reasons: it is a way of saying that competitive societies can never understand the world. But although the mere stating of this doctrine excites an immediate wish to demolish it, we shall postpone this pleasure to a later page. What I have in mind right here is rather this: that Protagoras's social intentions had a profound effect upon his theory of physical nature and of human knowledge. In order that some people at least might have success in the singular, he was willing to argue that no one can have knowledge in the universal.

It was the first great demonstration that the historian of nature is himself conditioned by nature and by history. The galling spur which Protagoras thus drove into her flanks made philosophy run for a long time through areas immediately human: the nature of society, the psychology of man, the possibility of knowledge. During this headlong journey Socrates died and Plato flowered and Aristotle instructed Alexander the Great. But ever afterwards it was only by reason of fear or enfeebled interest, and then only with much hopeful apologetics, that philosophy allowed itself to shrink from the problem of man and his place in nature.

We have said, in the first chapter, that this problem is philosophy's, that other disciplines are afraid of it and are in any case inadequate to it. I suppose that the positivists, who think that philosophy is only the "critique of language," and the pragmatists, who think it is a technique for quick profits, will consider this doctrine impossibly grandiose. Nevertheless, they have against them both the weight of tradition and the actual struggle of mankind toward a solution of the problem. There is, to be sure, no purely logical rule which can enforce acceptance of a definition. But if the positivists and pragmatists continue to identify philosophy with the fragments they have of it, the rest of mankind, who possess the whole garment, can afford to chuckle in their own totalitarian wealth.

Error and Society

The method of doing anything follows from the nature of the thing to be done. For example, the nature of the game called golf is such that you cannot play it with rackets or on skates. In fact, its nature is such that there is some question whether "playing" is what anybody does with it.

Similarly, if you think that philosophy is the analysis of language, then you won't philosophize about man's place in nature. You will behave more like a grammarian; and if you are not concerned to "settle *hoti's* business," that will only be because you are dealing with the living language of your customary use. But beyond the theory of language your philosophizing will not go. You will leave the description of the world to the scientists of the world, and you will be interested in politics only as a rich source of linguistic confusion.

If, however, you consider philosophy to be a union of theory and practice aimed at solving the problem of man's place in nature, then you will get a radically different method of philosophizing. Linguistic matters become, not primary, but incidental, to be treated whenever (but only whenever) their relevance is manifest. You will be concerned to settle, not *hoti's* business, but society's. As for the scientists and their descriptions of the world, you will want to know what they say and you will feel that they need to know what you say.

This view is much fortified by the fact that society has always been, and remains, a great source of philosophical error. I imagine that nearly all thinkers are aware of this. Some of them, indeed, feel it with such acuteness as to think that any contact with society sullies the web of speculation. At best, this timorous withdrawal has whatever virtue there may be in chastity untempted or virtue untried. At worst, it is an infantile blindness which supposes that evil, if unseen, does not exist. No thinker will ever know the full range of possible error unless he looks at society to see those errors rising upon him.

He will need to look carefully, because the rise of error has all the subtlety of infection. Living in a society requires that we share many things, among them *ideas*. The sharing begins at an age when we are totally uncritical, when we have not the faintest notion that there are such things as fancy or deceit. Our critical faculty, even when developed, is not drawn to all the errors which have sifted in; probably, indeed, there is not time to scan them all. And thus it happens, as Locke memorably said, that "doctrines that have been derived from no better original than the superstition of a nurse, or the authority of an old woman, may, by length of time and consent of neighbors, grow up to the dignity of *principles* in religion or morality." [1]

Such is the power of cultural lag, an enormous inertia of ideas which confers no benefit upon philosophy except to make it solemn. All its other effects have the form either of sloth or of paralysis. For example, criticism, awakening, fastens first upon details, but the generalization, the larger view, the ampler doctrine, which we took into our heads (according to tradition) at the maternal knee, never gets examined, never suffers the

[1] *Essay concerning Human Understanding*, Bk. I, Chap. II, Sec. 22. Locke's italics.

sweep of our happy broom. Then the doctrine, untouched by criticism, survives as a lure for facts we may later observe, which are thereupon entrapped like ants in a patch of glue. By dint of rationalizing, we can get a whole philosophy out of one of these persistent fossils; and since, fundamentally, we never look at things in any other way, this philosophy will seem to us to "work."

One of the spectacular instances of this phenomenon is the survival throughout modern philosophy, despite tremendous assaults, of the concept of the supernatural — that is to say, of beings or of forces "spiritual" in character and beyond space and time. Every great Western thinker since Descartes has destroyed some portion of the supernatural, but except for a few thorough materialists no thinker has succeeded in getting rid of it all. There are many social reasons for this, and among them the usefulness of the concept to ruling classes would, I suppose, rank high. But what I have in mind here is the fact that, in our tradition, no thinker, great or small, has avoided indoctrination by the concept before he was in the least able to examine it. Forgetfulness of origin, plus "length of time and consent of neighbors," leaves him with a point of view which he can lose hardly more readily than the lines of his hands.

Besides cultural lag, there is a perhaps more terrible force: the direct pressure of economic position. I do not here mean the obvious and degrading sale of ideas for income or the surrender of them before threats, but rather a custom or even a tradition to the effect that one ought to have the sort of ideas, as one has the sort of clothes, befitting one's income. The existence of this tradition is as plain as its content is astounding. The fact, for example, that I live in a pleasant house does not, I regret to say, alter in the least the condition of colonial peoples, and the truth of statements about them is in no way affected by the kind of house I live in. Despite all this, I have been told from time to time that I "ought" to consider colonial peoples as unable to be helped or unworthy to be helped or as eagerly awaiting Western enlightenment.

Now, unless I misread current history, none of these three statements about colonial peoples is true. But if, for reasons of income, I am required to accept those statements, then either I must accustom myself to believing false statements or I must adopt some method of distortion which will enable me, despite all evidence, to think those statements true. The first procedure is too cynical for wide adoption; it exists chiefly among journalists, who find it a professional necessity. The second, regrettably enough, calls upon past philosophy for a supply of arguments in favor of mystical ecstasy, pragmatic opportunism, or a lucky incidence of doubt. Cradled in his income of five thousand dollars a year, the sophisticate murmurs, "What can one know about colonial peoples when one knows so little about *anything?*" This is why the life of the intellect is too serious a thing to be left to the intellectuals.

The view that income determines knowledge is what one may call the suburban theory of truth. There are forms of it, however, much subtler than any I have yet described. As I look back over my years of study and of teaching, it seems to me that I learned something over and above fact and theory and method, something which might go by the name of "accommodation." That is to say, I learned not only philosophical theories and whose the theories were; I learned also that some of them (or some parts of them) were dangerous in the sense that they would bring penalties upon any man who openly professed them. Now, the wish to avoid penalties is very natural, and amounts in the last analysis to the wish to eat and to associate comfortably with one's fellows. At the same time, the educational system is such that a student does learn a certain number of true statements and with them the methods for discovering more.

Now, if true statements and statements free of penalty were one and the same, what I mean by accommodation would not occur. Unfortunately, however, history is full of penalized statements which were nevertheless true. I should suppose that everyone is at least dimly aware of this fact within his own experience. Accordingly, a student, learning how to determine truth, and learning also what statements are penalized, learns a third technique as well: he learns to accommodate truth to safety, to produce, that is to say, statements which are enough like reality to be credible but not enough like it to be unsafe. By this means he is able (or thinks himself able) to serve truth and self-protection at the same time.

The mechanism of accommodation does not ordinarily operate at a very high level of consciousness, because the dishonesty of it would then be intolerably plain. And as the technique becomes, through practice, a settled skill, these operations sink downward into the unconscious, until all feeling of a collision between truth and safety is lost. Thus trained, the thinker never utters any statement which would make him either an ass or a heretic. This skill is capable of astonishing refinement, and I have come to suspect that the philosophies dominant in our culture were arrived at in this way. For if, placing before them the great historical questions of our epoch, you ask, "Are you for capitalism?" they will reply, "Yes, with some reservations"; but if you ask, "Are you for communism?" they will reply unequivocally, "No."

A Case in Point

There is no great difficulty in recognizing when a philosophy has accommodated truth to safety, and indeed the enterprise offers all the adventurous pleasures of detection. The telltale fact is one of two sorts: either the theory does not draw certain consequences which, nevertheless, it logically contains, or it draws consequences which logically it does not contain. In the first case there is an abrupt halt just before the penalized statements are reached; in the second, conformist statements intervene without any basis in the theory's own assumptions. The logical imperfection of the theory shows that something other than science has been at work in the formation of it.

Let us now consider a case in point. The October, 1951, issue of a magazine called *The Humanist* has an article showing what happens when you ask intellectuals to state their personal philosophies. The writer, Mr. Warren Allen Smith, had confronted various authors with seven types of Humanism, and had invited them to say which of the seven they subscribed to. (One of the types was called "Communistic Humanism," and, as may be imagined, Mr. Smith got no candidates for that.) The replies were filled with protests about "labeling" and with ritual gestures toward semantics; but a certain number were willing to subscribe to Mr. Smith's own "Naturalistic Humanism," and the rest were willing to take it as a bona fide philosophy.

Now, here is what Naturalistic Humanism is, according to Mr. Smith's definition of it:

. . . an eclectic set of beliefs born of the modern scientific age and centered upon a faith in the supreme value and self-perfectibility of human personality; differs from Theistic Humanism by its rejection of any form of supernaturalism, from Atheistic Humanism by its optimism and relative agnosticism rather than absolute atheism, and from Communistic Humanism by its opposition to any beliefs not founded upon the freedom and significance of the individual.[2]

Among the authors who, according to Mr. Smith, have signified agreement with this view are Conrad Aiken, Van Wyck Brooks, Stuart Chase, Henry Hazlitt, Granville Hicks, Max Lerner, John Dewey, James T. Farrell, and Julian Huxley. Walter Lippmann, Archibald MacLeish, Thomas Mann, Bertrand Russell, and George Santayana appear as "associates" of the view — which is to say that they like the view but not the name. At all events, the lists show that Naturalistic Humanism is a theory not restricted to professional philosophers but alive and active among the intelligentsia generally.

As one reads Mr. Smith's definition of Naturalistic

[2] *The Humanist*, Vol. XI, No. 5, p. 194.

Humanism, the most noticeable impression is of pious and honorific terminology. All the great shibboleths of middle-class philosophy are there: "eclectic," "modern scientific age," "faith," "human personality," "freedom and significance of the individual." These terms have two sets of opposites. If you take "systematic" instead of "eclectic," "medieval religious age" instead of "modern scientific age," "revelation" instead of "faith," "human soul" instead of "human personality," and "subordination" instead of "freedom and significance of the individual," you have the characteristic concepts of feudal ideology which the middle class overthrew during the seventeenth and eighteenth centuries. Now, if you take as a second list "systematic," "age of human control over both physical nature and society," "science," "human organism," and "fulfillment of man's needs by co-operative social life," you get the characteristic socialist concepts which, it appears, are in turn to replace middle-class philosophy.

There is a nice dialectic in these relations, for the last set of concepts has obviously drawn upon the other two and has succeeded in unifying the material thus gathered. But the concepts expressive of Naturalistic Humanism, couched as they are in terms so honorific as to be commonplaces of advertising and public relations,[3] show a resolute wish to be neither feudal nor socialist. The last part of Mr. Smith's definition rejects

[3] In capitalist society nobody advertises a brand of toothbrush as "ordained by God," and in socialist society there wouldn't be any *brands* of toothbrushes to be advertised. Consequently, advertisers generally describe their clients' toothbrushes as products of the modern scientific age or as contributing to the freedom and significance of the individual.

"Communistic Naturalism" not only because of immediate political pressures but because the two theories represent different historical epochs and antagonistic social systems.

This conflict of epochs and systems is remarkably expressed in the treatment which Mr. Smith's definition gives to medieval philosophy. There is in that treatment a charming mixture of contempt and tenderness toward a defeated foe. There is also a logical contradiction. We are told that Naturalistic Humanism rejects "any form of supernaturalism"; we are also told that it is marked by "relative agnosticism rather than absolute atheism." This last suggests a maiden's wish to be relatively virginal rather than absolutely pregnant: it is easy to wish and difficult to achieve. For if you are only *relatively* agnostic, you may on occasion lapse from the agnostic norm; and if you lapse from the agnostic norm, you must do so either into theism or atheism. Or, to release these doctrines from their cloaking of nouns, we may say: If you are not quite sure that the existence of God cannot be proved or disproved, then there remains a possibility of your believing that it *can* be proved or disproved.

Now, the concept of God is the concept of a supernatural being. If, therefore, Naturalistic Humanism rejects "any form of supernaturalism," it must also reject the concept of God. It must hold, in fact, that there isn't any God. But this is "absolute atheism," to which, however, "relative agnosticism" has been preferred. Plainly we can't maintain all of these doctrines at once — not logically, at any rate. If we do maintain them all at once, the reasons will have to lie outside of logic.

And so they do. They lie, to be precise, in history. The rejection of "any form of supernaturalism" is a twentieth-century echo of middle-class thought in its revolutionary period, when the materialist implications which inquisitors correctly discerned in it rose for a time to the surface. But these implications were thrust back into the depths again as soon as they threatened to expose the nature of middle-class rule. Accordingly, the sophisticates, who cannot endure to be obviously deceived or to be obviously heretical, must accept "relative agnosticism rather than absolute atheism." They take the supernatural not too seriously and at the same time not too frivolously. They will not be laughed at by their friends or arrested by the police.

Thus in Naturalistic Humanism the accommodation is plain. It bears indelibly the print of its bourgeois origins, even to the point of tolerating internal inconsistency in order that it may nevertheless say the "right" things. Its forgetfulness of any economic source, its belief that it sprang full-panoplied from a corporation of enlightened minds, is the most characteristic thing of all. For here we see an accommodation so skillful that the collision between truth and safety has occurred without shock and, it would seem, without notice.

How Shall We Philosophize?

If philosophy is an undertaking to formulate true statements about the world, Naturalistic Humanism is evidently not the way to philosophize. The moral is that philosophers, to be on guard against error, must be constantly aware how society affects them. Otherwise they cannot catch the odious little compromises by which safety modifies truth. They may, and do, fancy that they are describing nature; but nature, it is clear, they cannot describe unless they simultaneously consider man's place in it.

What, then, should be the true method of philosophizing? Ideally, it would be one capable of describing things as they are, without distortion by fear or interest or the narrowness of our own faculties. I suppose that no philosopher nor any group of philosophers is quite able to do this; consequently, as Descartes observed, "notwithstanding the supreme goodness of God, there is falsity in my judgments." But error, if it cannot be entirely avoided, can be minimized. This possibility transforms itself into a duty for every philosopher and, indeed, for every man.

What is thus required is a point of view centered in human society and at the same time capable of vision throughout the whole stretch of environing nature. Further, the unity of theory and practice requires that the observations thus made shall be organized and interpreted in relation with human needs. Solitude, aloofness, and pessimism are thus excluded; the philosophic effort itself assumes society and invokes fulfillment. If success were impossible, we should not have risen to these heights in the first place.

Let us, accordingly, take ourselves as statesmen in the best sense, that is to say, as persons concerned with making mankind safe and happy in the universe it inhabits. To us, in such a case, all things grow relevant and useful. The sterility of thought without action and the blindness of action without thought will trouble us no longer. The biases of selfhood or of class privilege will be canceled because we are thinking of mankind, and physical nature will at last appear a garden rather than a place of terror or of retreat.[4]

If these assertions are true, we must expect to find the main problems of philosophy — reality, knowledge, value — arising directly within the lives of people themselves. Any life will serve as a source and an example, but I shall propose one from the submerged multitudes in our society. If the Gospel of St. Matthew is right in taking as the test of ethical achievement our behavior toward "the least of these our brethren," we may suppose that the full meaning and power of philosophy are displayed there too. For a willingness to right the wrongs which the exploited suffer is the best proof of our intent to think universally, and a philosophy which can thus assist the humble is able to help all.

Let us therefore consider a certain woman, a fellow citizen of ours, to whom in our canting way we have ascribed the forms of freedom and equality while canceling the substance of both. Her name is Mrs. Isaiah Nixon. She is a Negro, and she lives in Georgia.

On the eighth of September, 1948, her husband appeared at the polls and voted in the primary election. That same evening, a group of white men visited his farm. When they left, Isaiah Nixon lay dead of a bullet which had ended his life, his liberty, and his pursuit of happiness.

---

[4] Ignorance unrelieved makes nature seem terrible. Ignorance sophisticated by science makes it seem a welcome refuge, e.g., in Bertrand Russell's remark, "I think, on the whole, that the nonhuman part of the cosmos is much more interesting and satisfactory than the human part." (*The Humanist*, Vol. XI, No. 5, p. 199.)

Here is Mrs. Nixon's story, composed into a single narrative from answers to several questions:

Isaiah Nixon was my husband. He was a good husband. It was on September the eighth he got shot, on votin' day.

Isaiah went to the poll. He voted, and after he got through he came on back to the house and set down and listened to the radio. He played with the baby for a while, and then he got up and he went in the kitchen, and he had his supper. He had peas and bread, biscuits and meat, and sweet potatoes for supper.

And I went out on the porch and I saw a car coming, and the car drove up, and they asked was he at home, and I told them yes, and they asked him had he saw Evan Johnson, and he told them no, sir, he hadn't saw him that day, and he said, "Well, step down a minute," and he went to the fence where he was, and he throwed the gun — the pistol — on him, and come back toward the house, and the white boy, he — And they just shot him down. That's all I can tell you.

I don't know why they'd shoot him no less'n it was 'cause he voted. 'Cause he had a rights to vote. He hadn't done anything to them. He hadn't had no words that I know of.

I sit and think of him when night come.

I have six children. I don't hardly know how to take care of them by myself. I needs help with them.

I wish I could explain just why and know just why we have such a hard time in this part of the country.[5]

"Explain just why and know just why": language so guileless and direct mocks my poor pretentious phrases. When I tried to state the content of philosophy, I said it was Man and His Place in Nature, and I was driven

---

[5] This eloquent narrative, spoken by Mrs. Nixon herself, was recorded on tape by the Progressive Party and used by them in their Election Eve broadcast in 1948. I am indebted to the national office of the Progressive Party for the written text.

to adorn even this with capitals. Mrs. Nixon, however, shows that what I meant was rather this: the task of philosophy is to explain just why and know just why we have such a hard time in this part of the country. And "this part of the country" is anywhere in the world that times and lives are hard.

What is involved in the explaining and the knowing that Mrs. Nixon speaks of? Remembering our wish not to sever theory from practice, we shall approach this subject, not as mere observers, but as people who propose to alter circumstances, to remove the hardness from the times, the anguish and frustration from the lives. This being the aim, we shall find that we have to make certain philosophical assumptions:

(1) We have to assume that change is real, that it is indeed a basic fact about the world. If change were unreal and illusory, then any alteration we might propose would be unreal and illusory also. Or, if change had the status of a minor fact, the "real" reality being changeless, then we could help Mrs. Nixon only in trifling ways.

(2) We must next assume that change is controllable by men. For even though change is a basic fact, it may be of a sort we cannot manage. That is to say, it may be the case that events *always* happen just as they happen, regardless of what *we* do; or even that our doings are small predetermined noises in the great machine. Therefore it doesn't suffice for change to be real; it must be controllable also.

(3) We must assume that we possess, as human beings, the means for controlling change. That instrument, it seems obvious enough, is knowledge. We are

assuming, therefore, that knowledge is possible and that we do not pass through the world in total ignorance of what it contains.

(4) However, in order to use knowledge, we must know when we have got it. All knowing consists in a choice among statements, a decision as to which, if any, of them is true. For this we need a method of inquiry and a standard of truth. The possibility of these is therefore also assumed.

(5) Lastly, we need to know the purpose and goal toward which change ought to be directed. By "ought" I mean something much more rigorous than the statement "John wants this and is therefore doing that." I mean that we are to assume the possibility of a rule for choosing right actions, just as we assumed the possibility of a rule for choosing true statements. Otherwise there will be no such thing as marshaling power on behalf of righteousness. There will only be a marshaling of consent on behalf of power.

These are the five philosophical assumptions which anyone must make who wants to help Mrs. Nixon. Obviously they commit us to quite definite views upon the great traditional problems of philosophy. Our task is to show whether the assumptions are warranted — whether, that is to say, we can get affirmative answers to the following questions:

Is change real?
Is change controllable?
Can we have knowledge?
Can we know when we have it?
Can we know what ought to be done with the knowledge we have?

If the answer to any of these questions is No, then we can do nothing good for Mrs. Nixon or for anybody else in the whole wide world.

Permanence versus Change

The rest of this book will be an attempt to answer all five of these questions in the affirmative. It will be a Yea, which, if not Everlasting like Carlyle's, will at any rate be uttered once. That single, encouraging syllable is worth uttering; the more so, indeed, because it can be sustained by argument. Moreover, my convictions about the course of history are such as to leave me in no doubt that mankind, in actually working upon these problems, will achieve a very evident success. While linguists whisper and logicians cough, history with less gentle noises settles everything.

Doubtless for those of us who live within the process the movement toward solution follows a path with no very visible end; in a sense, it embraces the whole future development of man. But in the mid-twentieth century we stand on no small eminence, and our vision is obscured more by storms than by low altitude. Behind us lies the conquest — partial, to be sure — of physical nature, together with the science and philosophy which have brought all that about. Before us lies our conquest of society, that is to say, our final mastery over ourselves. And beyond that, an endless Eden in which no fruits are fatal and no men are doomed to fall.

I suppose this is a kind of humanism, but one not locked within the present epoch nor chained to any past. It proposes to discuss philosophy with a good deal of attention to inherited ideas and to current discussions, but also, so far as may be, in the light of predictable developments. The five questions to which, as we have seen, Mrs. Nixon's more general question gives rise, fall into three groups: (1) the problem of change, (2) the problem of knowledge, and (3) the problem of moral value. In the pages which follow, each of these is the subject of a separate section. They could appear in any order, for philosophy is a sort of globe which can be entered at any point upon its surface. However, the sequence I have chosen, from cosmology to ethics, seems to move toward climax, and in any event I think it useless to speculate upon ethics before one first knows that change for the better is possible. In other words, if my conclusions in the first section are erroneous, the rest of the book may be thrown away.

We shall begin, then, with the reality and nature of change, and I think that something should be said here about the problem generally, before we pass to a discussion of existing theories in the next three chapters. Perhaps it will be best to start with certain difficulties which people find in the notion that the world is constantly changing.

One of these difficulties is that when we say the world is constantly changing, we are saying (by logical equivalence) that nothing lasts — except, perhaps, the world itself. To the extent that we are fond of variety, this is a very agreeable notion: it suggests that the universe, whatever else it is, will not be boring. But, as a matter of fact, although we love variety, we also love to have some things last. "Establish Thou the work of our hands upon us!" cried the Psalmist, who knew well the power of that wish. The yearning for security, which I sometimes think is the bottom of all human wishes, inevitably connects itself with the idea of something permanent. If you happen not to be aware of this fact, just fancy to yourself the horrors of a universe of pure chance in which nothing occurs but the unexpected.

The lure which permanence exerts upon our wish for security is much strengthened by the fact that the world of change, as we daily experience it, is for most people not remarkably satisfying. Condemned to poverty, disease, and war, they have learned from the slow or giddy movements of their lives a lesson (or so they take it to be) that time blights every hope. Accordingly, change comes to bear a bad reputation. The feeling grows that this universe of space and time, this fevered process, this resistless tide, is somehow an enemy to man. And since complete pessimism is reserved to people (like Schopenhauer) who don't really suffer *very* much, the more usual human reply is to conceive a universe beyond space and time, in which some entities exist fixed and unalterable.

The most familiar examples of this notion will be found in the religious literatures, where the steadfast entities tend to be persons rather like ourselves. It is possible, however, to fill these altitudes with abstractions: truth, beauty, and goodness may dwell there like Beatrice in the celestial rose. Whatever the entities may be, they are one and all projections of desires which have been denied or cruelly teased in the world of time. In their projected form they are, as we like to say, "eternal," and they enable us to drop a spiteful hint to time that it has cheated us in vain. The turbulence of its waters was only foam after all, and had no power upon the geography of our love.

But we have other motives for cultivating permanence. The struggles within mankind are now so sharp and the decisions so critical that we find it less and less congenial to proceed by pragmatic trial and error. The old-fashioned "self-made man," whose ignorance of sociology was so great that he never could understand his own success, always supposed it due to a faculty he possessed for hitting on schemes that worked. It is difficult, however, to bet your life on such a mysterious talent. When the great decisions come, you begin to look for something fixed on which to base them. You begin to feel that there is a law above our law, a truth above our truth; and in this language the praiseful preposition "above" signifies an order of existence permanent and only approximated within space and time.

In this way and in terms which I think everyone will find familiar, the unhappy divorce between wish and fulfillment gets transformed into a divorce between two worlds. What began as sociology has become metaphysics, and at this point the professional philosophers (if we may so style them) enter upon their task of rendering the whole notion coherent. This particular notion has been under study since Plato's first brilliant conception of it. A long time ago, but even so not nearly as long as the social agony of our race.

Precisely because these transcendental philosophies, invoking eternal worlds and eternal beings, do express the yearnings of mankind, they deserve respectful treatment. The problem is to destroy the metaphysical ground without harming the values imbedded in it. The operation is delicate — and necessary. For, so long as values are thought to be eternally realized in this illusory way, they tend not to get realized in our actual lives. Perhaps the most sublime pathos there is will be found in those multitudes upon multitudes of men who seldom possessed what they wanted, yet imagined value to have lain with them all the while.

Thus men, impelled by struggle and frustration and even (it has sometimes seemed) by logic, have set up permanence as a barrier to change. The effort is understandable. It is, however, mistaken, and in the hands of servile theorists it enables rulers to prevent just those changes which would confer most benefit.

Seen thus, the question takes on somewhat the warmth and interest of a personal, practical problem. I have thought that we might divide discussion of it into three stages: (1) a view, once dominant and still formidable, which holds the universe to be in fact immutable, with change as a sort of misleading shadow upon its surface; (2) a counterview, still dominant but not so formidable, which holds the universe to be a very loosely organized flux; and (3) a view which holds that permanence and change require each other, and that these two principles, taken together, will suffice to describe the universe as it is. The first of these views is our inheritance from certain English and American philosophers who derived from Hegel and from the German idealism generally. The second, a revolt against the first, is the pragmatic philosophy which originated in James but which has since passed (with increase of respectability) into the main stream of empiricism. The third owes something to Whitehead in the twentieth century, but much more to Marx and Engels in the nineteenth.

The universe, so long as we live in it, is *our* universe. Let us take it, for a time, as a sort of parish, and inquire what really does occur "in this part of the country."

THIS PART OF THE COUNTRY

CHAPTER III PERMANENCE
                 WITHOUT CHANGE

HEGEL'S PHILOSOPHICAL CHILDREN, one of whom rather doubted the derivation,[1] were lordly and a little unclear. For almost a century, however, they bestrode the Western earth like colossi, while lesser intellects crept about their feet seeking clay. Even as late as twenty-five years ago, a beginner might come (as I did) under the spell of Hegelian tutors, and learn from them "the consummation and the poet's dream." Then, about the time of the Great Depression, the earth shook fatefully, the giants vanished, and of that vast edifice which had been the Absolute there remained only a solemn, granite ruin.

The lordliness of Hegel's children had some snobbery but more charm. They measured themselves with satisfaction against the cosmos, but they were quite willing

[1] This was Francis Herbert Bradley, who wrote, "For Hegel himself, assuredly I think him a great philosopher; but I never could have called myself an Hegelian, partly because I can not say that I have mastered his system, and partly because I could not accept what seems his main principle, or at least a part of that principle. I have no wish to conceal how much I owe to his writings; but I will leave it to those who can judge better than myself, to fix the limits within which I have followed him." (*The Principles of Logic*, London, Oxford University Press, 1922, Vol. I, p. x.)

that any man should do so. In the same way, their lack of clarity caused some exasperation but more sympathy. For the secrets of existence, whose keepers they fancied they were, are not very readily told, and the effort to say everything about the universe has the sad result of making it difficult to say anything. This was their burden, and it must be very commonly borne.

The giants, moreover, were genuine. They had stature and could see afar. Empiricists, their familiar critics, were (and still are) wont to sit like children on a beach, eyes bright with narrow vision, nostrils quivering with immediate sensation, holding up now a pebble (Look, how round!) and now a shell (Feel, how smooth!). But the Hegelians, from their height of head, saw the whole immense horizon, and the arching clouds, and beyond the clouds the vault of sky, and beyond the sky a great many things that were not there. It was rather a new habit of thought for Anglo-Saxons, and you may be interested to know how it occurred.

Hegel and the Anglo-Saxons

"To an Englishman," wrote Philip Guedalla in his Oxford-Union manner, "his island is a piece of land entirely surrounded by foreigners." In the nineteenth century the foreigners thought so, too; and, reflecting upon the cultural insularity of Englishmen, they found some recompense for the fact that, commercially, those same Englishmen were surrounding everything else.

The French philosopher Victor Cousin remarked in 1828 that "England is not destitute of inventions, but history declares that she does not possess that power of generalization and deduction which alone is able to push an idea or a principle to its entire development." [2] This was a fairly rash judgment about a country which had produced Newton and was even then producing Darwin.

Nevertheless, England was culturally insular, as befits a nation with an expanding economy and with shrinking doubt as to its own excellence. By 1842, Tennyson had indulged the dreadful idea of chucking Europe in favor of life among the South Sea Islanders, there to "rear my dusky race." The spectacle of the future laureate turning primitive is one of the most comic fancies imaginable. Nothing of the sort could possibly happen: history is not so blithe. But Tennyson's recoil from the notion seems a trifle overshocked: "I the heir of all the ages in the foremost files of time." In other words, history had culminated in Victorian England, and any further progress would have to be in degree and not in kind.

Another barrier to the entrance of ideas was the preoccupation with Anglican orthodoxy. This was the epoch when Mr. Gladstone declined to employ a servant girl until he had first discovered her opinion about the relation of Church and State, when a certain Robert Blakely stood (as he said of German thought) "aghast in amazement at the audacity and folly which gives utterance to doctrines so denuded of every particle of scriptural authority and common sense," [3] when Dr. Arnold set before the dazzled youth of Rugby the double ideal

[2] Quoted in J. H. Muirhead, *The Platonic Tradition in Anglo-Saxon Philosophy*, New York, Macmillan, 1931, p. 149.
[3] Quoted by Muirhead, *op. cit.*, p. 151.

of being a Christian and an Englishman. A point of view so narrowly theological could not willingly admit a philosophy in which the transcendence of religion was itself to be transcended.

Nor was the philosophical climate of England more inviting. Sir William Hamilton, who was acquainted with Kantian thought, drew from it, not illegitimately, the conclusion that ultimate reality is unknowable. Herbert Spencer was, somewhat later and on different grounds, to reach a similar view. George Henry Lewes (more identifiable, perhaps, as Mr. George Eliot) understood Hume to have ended philosophical speculation, and accordingly held that "the interest in philosophy has become purely historical." [4] These men seem, in Engels' phrase, to have translated their ignorance into Greek and called it Agnosticism.

The inroads of Hegelianism were therefore slow, but they were also steady. Earlier in the century, Coleridge had intimated the fact that he could read German. Thirty years later, Carlyle was trumpeting a similar talent from Craigenputtock to Chelsea. Indeed, Carlyle sometimes wrote English as if it were German. What thus entered English thought was not in fact Hegelian doctrine but the German idealism from which it sprang and of which it was the chief glory. But Carlyle had breached the island defenses; and when Hutchison Stirling, after a sojourn at Heidelberg, returned to write during eight long years *The Secret of Hegel,* all the defenses went down.

In America, where philosophical thought had been imitative for more than a century, the situation was ap-

[4] Quoted by Muirhead, *loc. cit.*

propriately similar. Emerson was serving as a paler, less boisterous Carlyle, the other Concord wits were stirred with various yearnings, and Margaret Fuller, after a season of meditation, accepted the universe.

There were even the same theological barriers. Harvard looked darkly upon the heretical implications of transcendentalism, Yale expelled the Kantian philosophy as morally subversive, and the trustees of Princeton announced a hopeful intent "to make this institution an asylum for pious youth in this day of general and lamentable depravity."[5] Fate can be unkind to the noblest wishes. The present philosophy building at Harvard is called Emerson Hall; Yale has at least one Kantian on the philosophy staff; and, unless the number of pious youth has increased since my day of general and lamentable depravity, Princeton has offered its asylum in vain.

When centers of learning prove stubborn to new ideas, the ideas simply grow up around the centers. Accordingly, it is pleasant to know that Hegelianism came to America by way of the old frontier. In the 1850's, a remarkable immigrant, Henry C. Brockmeyer, after an unsatisfactory taste of American college life, went out into the Missouri backwoods, seeking wisdom direct from nature. He thus became the only transcendentalist to obey that theory's best injunction; and, reward crowning endeavor, he arrived at last in St. Louis.

Here, in "a Teutonic city of the radical type," he met William Torrey Harris, in whom he aroused so much zeal for the cause as to produce "the most devoted

[5] Quoted from an early Princeton catalogue, by Muirhead, *op. cit.*, p. 317.

Hegelian propagandist that ever lived." This reunion, upon foreign soil, of German culture, German beer, and German philosophy was irresistible. The fact that Harris, after importing a copy of Hegel's *Greater Logic,* found he could not understand anything beyond the first part of the first volume did little to dampen enthusiasm. For, in the nineteenth century, thoughts sufficed provided they were spacious, and clarity was the hobgoblin of little minds.

Accordingly, Harris, taking as much of Hegel as Hegel would let him take, went on to found the *Journal of Speculative Philosophy,* with a laudable desire "to elevate the tone of American thinking," as the "Address to the Reader" tells us. This desire was in part fulfilled, but chiefly by the labors of a man born yet farther away from the New England enlightenment: Josiah Royce, the Californian.

Royce had had a brilliant career as a student, and when he settled into an instructorship in rhetoric and logic at the University of California he must have felt himself sadly isolated. At any rate, in 1882, he seized the opportunity to come to Harvard as a substitute for William James, then on sabbatical leave. In doing so, he resigned a permanent position, and brought his wife and baby three thousand miles — all for a one-year appointment at a salary of one thousand dollars.

George Herbert Palmer, who records the story from first hand,[6] observes of the California environment that "things of the mind were little regarded by seekers for gold." The economic conditions offered Royce will show

[6] In a memoir of Royce, in *Contemporary Idealism in America,* edited by Clifford Barrett, New York, Macmillan, 1932, p. 6.

what value American universities place upon things of the mind.

Royce, indeed, would have had to leave Harvard after the first year, except that Palmer himself went on sabbatical and a substitute was again required. After the second year, Royce was indispensable. There was a brush with the University officials and the loss of a thousand dollar fee, when he declined to give the Lowell Lectures because these demanded assent to certain religious doctrines. But from 1885 onward, when he had published *The Religious Aspect of Philosophy*, the Harvard students and the American illuminati in general listened to a ceaseless movement of remarkable prose, very much like hardtack in taste and nourishment. The students, says Palmer, "knew that something big was going on above them, and were all duly elevated." [7]

In physical aspect, Royce seems to have been more striking than the vague contours of most philosophers permit them to be. "He had," says a recent writer, in words which I take to be no flattery to the profession, "a round ungainly figure, with slight muscular development, but his great dome-like head was impressive, and all who saw him knew him at once to be a philosopher." [8] *Et vera incessu* . . . The true philosopher reveals himself in his dome. Ah, Virgil!

Such were the early adventures of Hegelianism when it had new worlds to conquer. Some men whom it mastered never mastered it, some whom it mastered thought they did not agree with it, and some agreed with

[7] *Ibid.*, p. 7.
[8] William Kelley Wright, *A History of Modern Philosophy*, New York, Macmillan, 1944, p. 486.

it not knowing what they thought. For an entire epoch the universe lodged comfortably within the finite human mind, and gave thereby the impression of mind stretched to infinity. And men who found in Nature little to love became the familiars of systems, categories, and cosmic spirits. It was the last time that middle-class philosophy was sure of the world.

The Absolute

We have said that the Hegelians, in looking about the universe, saw many things which were not there. One of these, I rather think, was the Absolute. The concept, like the name, is formidable. It is, however, extremely interesting in itself, and it (or something very like it) is what you will have to hold if you want to deny the reality of change. Perhaps not very many people would wish to go this far, but, as we observed in the previous chapter, a lot of them are quite willing to reduce the importance of change in the universe. They tend, therefore, toward the Hegelian view as toward a limit, and that limit makes much plainer the view they actually have.

Suppose, then, that you abolish change. You abolish time with it. Time being gone, what about space? Space without time would have no locomotion in it; it would be a dead solid, like a cannon ball. Parmenides held this view, *circa* 500 B.C., but except for his immediate disciples no one has since found it credible. Apparently, if time goes, space goes too. The altogether changeless universe, if there is any, is therefore "outside" or "beyond" space and time. There is some irony in the fact that you have to state this notion in spatial metaphors. What will such a universe be like? Well, it will be the sort of entity that space and time can be "in," while preserving for itself quite a different character. Take an analogy from common life. You won't think it strange to be told that you are a conscious person and are therefore aware of objects moving in space. You can think of the objects moving in space as a different sort of existence from the consciousness which is aware of them. In short, you can think of the objects as being in space and time, and of your consciousness as being "outside" space and time. One further thing you will notice is that your consciousness remains exactly the same consciousness, no matter how much the objects move.

Now, project this whole conception upon the universe. The result you get is that there exists a cosmic consciousness which stands to objects in the same relation that your own consciousness does. It is "outside" the objects, just as your consciousness is. It is always the same consciousness, just as yours is. Further, it is a safe repository for values, and this, of course, is true of your consciousness also. Call this cosmic consciousness the Absolute, and you have at last reached what the Hegelians intended. As Josiah Royce believed, the Absolute possesses all human knowledge, and possesses it in the way men do.[9]

And so in a sense it does. But the Absolute also knows things in a way we don't know them, namely, that it knows them all. This is a tremendous difference, for it means that we and the Absolute are as far apart as the

[9] See *The World and the Individual*, New York, Macmillan, 1901, Vol. II, p. 408.

limited and the unlimited, the finite and the infinite, the conditioned and the unconditioned. There is only one way you can know *everything*, and that is simultaneously. For example, to know a thing as a future fact is very different from knowing it as a present fact. Consequently, if there are some things which the Absolute knows only as future facts, then the Absolute doesn't know everything. But this result is contrary to our hypothesis. Therefore the Absolute, knowing everything, knows everything at once.

This gives us the universe we wanted, safe and sound. In all that is really real, change has been abolished; and the suspicion grows that it is some sort of illusion arising, as Hegelians loved to say, from "our finite point of view." This assertion, however, does not mean that our impressions of time and change are grossly erroneous, like a belief in ghosts. It means that time and change are "lesser" realities consequent upon our human limitations.

Now, as a matter of fact, some errors do result in just this way. In ancient times, for example, there were various fanciful conjectures about the earth's shape, all of them due to the fact that an observer had to occupy some point upon its surface. If you could, in those days, have looked down from the stratosphere, you would have seen the curvature and might then have guessed the global shape. But as an ordinary earth-bound observer, with not much more than immediate sensations to go by, you might easily have come to Anaximenes' conclusion that the earth is a disc floating on air.

You would have been mistaken, of course, but the fault would have lain in what you didn't know rather than in what you knew. Further knowledge would have led you toward the notion of a sphere. Similarly, so the Hegelians would say, change is what you ascribe to the world when you don't know the world well enough, when you don't know it with that complete and perfect knowledge with which the Absolute knows itself. You are not so much mistaken as misinformed.

Error is thus a simple lack of knowledge. The conception is very old, but it remained a favorite with the Hegelians. Royce, indeed, was able by the use of it to argue that the existence of error in men proves the existence of perfect knowledge in the Absolute. You need to suppose complete knowledge somewhere, in order to be conscious that you haven't got it all.

This conception of the universe as a system of things known — as a system, in fact, in which things exist only because they are known — reflects many of the motives of modern commercial society. There is, for one thing, an obvious influence of physical science: the Hegelians did all they could to endow human knowledge with certainty. There is in the concept an eternal resting place for values — not intended, however, as a burial ground. Above all, there is a warm, perhaps a pathetic, confidence in the perpetual stability of things; and this confidence lasted until the only permanence left in commercial society was permanence of crisis.

In the old days of great hope, however, feelings were different. The historical origins of philosophy could get lost and forgotten in a method of speculation by which states of existence were defined as states of being known and change was defined as superficial distortion of the real. Even the hymn of praise itself which Hegelians sang to the universe passed for mere description of fact. "It is not the business of philosophy," said Bernard Bosanquet, "to praise the universe or to exalt the satisfactions of goodness." [10] Which resembles much the saying among cats that it is not the business of philosophy to praise cream.

The fact is that Hegelians loved the universe, or, at any rate, what they took to be the universe. They loved it with an exaltation which in less sober men would have been rhapsodical. They loved it, indeed, so much that they were unwilling to miss a fragment of it, but kept it all real without loss or diminution by time. Their finite selves fed upon infinite substance; they turned the bread and wine of planetary existence into the body and blood of eternity. It was an unorthodox and stupendous miracle.

And they came upon the Western scene at just that moment when men appeared to have only two choices: the thin, calculated delights of the utilitarians or the thick, disordered woes of the pessimists. Now, as between a theory which holds that there can be no pleasure without arithmetic and a theory which holds that there can be no pleasure at all, the preference seems pale. One gets a sudden vision of Jeremy Bentham rushing in with charts, graphs, slide rules, and thermometers to decide the issue. But the Hegelians, whatever else they were, were robust, and drank in pleasure with such a natural draught as never to forget the source of ecstasy in the joy of it.

I suppose it was this which enabled them to look upon

[10] *The Value and Destiny of the Individual*, London, Macmillan, 1913, p. 327.

PERMANENCE WITHOUT CHANGE   57

the world's pain and evil with so sympathetic a calm. Those who suffer under evil take a different view, which may be struggle or resignation or despair. The Hegelians saw that under the world's curving periphery lay much jagged and lacerating metal; yet, as they were always quick to add, they also saw the sphere.

Such calm seems almost frozen, like the Absolute, but the Hegelians never spoke in icy metaphors. For them, rather, the universe was an eternal garden, thick with seeds and blossoms, with leaves maturing and leaves dead, and all the vast issue of generation and decay flung pell-mell over the cosmos. "The Absolute has no seasons," said Bradley in his memorable way, "but all at once bears its leaves, fruit, and blossoms." [11]

The universe contains everything that is, and there is nothing outside the universe. Nothing can be lost to the universe, for that would diminish its reality. Nothing can be added to the universe, for that would mean the universe had been imperfect. There is no before and after, no time, no change. And the spirit of Parmenides, surviving the death of empires and social systems, was heard again, crying, "Being is, and Not-Being is not: this is the sure path, for truth attends it."

It is difficult not to suppose that this love of the universe expressed also a vast satisfaction with existing society. Accordingly, Hegelian metaphysics is tied to a certain stage of social development; and, as that stage begins to vanish from history, the metaphysics generates reactionary implications. This fact is startingly demonstrated in the ideology by which the quasi-fascist gov-

[11] *Appearance and Reality*, London, Allen and Unwin, 1916, p. 500.

ernment of the Union of South Africa defends its oppression of the native peoples. Consider this passage from a news report in the *New York Times* for June 15, 1952:

JOHANNESBURG, South Africa, June 10 — An insight to the mentality of the ultra-nationalism that is the source of Prime Minister Daniel F. Malan's political strength in the Union of South Africa is given in an article in Inspan, Afrikaans business magazine, by Dr. D. N. Diederichs, prominent Nationalist member of Parliament and big business representative.

Dr. Diederichs, a former university professor and one of the principal philosophers of the Nationalist movement, poses this question:

"What do we mean by 'Afrikander'? What are the most noteworthy qualities of our people, as a people, that distinguish it from other people and make it an independent reality?"

The Great Trek into the interior a century ago was "a revelation of the real nature of the Afrikander," Dr. Diederichs declares. Speaking of the religious awareness and "individualism" of the Afrikander, he says:

"He saw himself as part of the Creation, but separate from the rest of the created world in that he carried with him a divine element. He saw himself as a link between eternity and the temporal, and therefore as something particular to itself and unique with its own reality and value.

"We who are Afrikanders are, in the first place, a people because we believe that we belong together, because we as a people, have been called to fulfil a God-given calling."

Dr. Diederichs devotes the second part of his article to an analysis of those "factors from outside influencing the Afrikander people today." The first of these is a lack of belief, he says.

PERMANENCE WITHOUT CHANGE

The second, he went on, was the drive toward equality that derived from a lack of belief.

"A world that loses all sense of the absolute must necessarily deny the hierarchy of values," he says, "and finally do away with all differences, discriminations, and dividing lines, which leads to equality, leveling, the destruction of that which distinguishes one from another.

"The trekker (Dutch pioneer) observed and maintained differences and lines of division. The divisions of day and night, summer and winter, rain and drought, black and white. But the world of today does not want that. The world of today is the world of the masses, of the average. In the mass everything is alike."

Thus the consequences of Hegelian theory are prodigious. Systematically deduced, they will affect every decision we can make and every method of inquiry we can employ. For example, the theory tells us that all problems and their solutions exist, fully known, in the Absolute. Our procedure, when confronted with a problem, will accordingly be to discover the solution which already exists, not to formulate any programs of active change. We get the curious sense that all conflicts have been previously resolved, all battles won.

There is a visible comfort in the theory, especially for times when not much can be done in a practical way. But I think we may doubt that Mrs. Nixon (for example) would prefer a past solution of her problems in the Absolute to a future solution of them in space and time. Indeed, Hegelian comfort seems a little ersatz. We can count its inadequacies, one by one, if we observe how the theory works itself out in ethics, in politics, and in the public life of one of its greatest exponents, Bernard Bosanquet.

Ethics

If you take ethics seriously (the phrase is a favorite among those philosophers who take it very seriously indeed), there are certain statements you must acknowledge as true. You must acknowledge, for example, that at any moment there are alternative courses of action among which people can choose, that the choices will differ in merit (some being better and others worse), that consequently there is a recognizable difference between good and evil, and that the recognition of this difference is not a matter of personal whim or even of social custom.

But, most of all, you will have to acknowledge that, if ethics is to mean anything, the actions based upon it must be able to change the world, making it what it would not otherwise have been. For if, despite your body of principles and the choices carefully based on them, your actions can effect no change and the world continues stubbornly as before, then principle, choice, and action are alike vain. Ethics might survive as a mental pastime exhilarating to some tastes, but it would have nothing to do with getting us around in the world.

Now, if, as Hegelians say, change is a "lesser" reality, and if ethics is concerned with change, it follows (tautologically, in fact) that ethics is concerned with a lesser reality. This conclusion is as alarming as it is inevitable. For who will bear the sweat and agony of choice in order to grasp something very like illusion? And if ethics is not a discipline by which we can reach the full reality, we seem to have an excellent reason for dispensing with ethics altogether. Yet in the doctrine there is a certain comfort. Our efforts at morality are not ordinarily such as to leave us dizzy with success. Where we but feebly gain, and others do not even try, and still others (it would seem) go militantly wrong, there is solace in the view that the universe covers all failures, lethargies, and evils with a goodness immaculate and divine. Pippa passes from Hegel to Royce: good's in the Absolute, all's right with the world.

This comfort, however, conceals a further problem, which is so old as to have an official name. It is the Problem of Evil, and it embarrasses those philosophies which, having pronounced the universe good, must then explain the presence of evil in it. For myself, I doubt that this can be done, and I think that what is required is a change in the philosophy. Nevertheless, we may look at the Hegelian efforts:

(1) One solution might be to say that evil, like pain and error, is part of "the hazards and hardships of finite self-hood." [12] Then the triumph of good over evil can be thought of as our transcending our finiteness. Of course we can't *really* do this, since finiteness is transcended only in and by the Absolute. But perhaps we can transcend approximately by widening the limits of our knowledge and by finding our place in "the whole world of achievements, habits, institutions in which the apparent individual finds some clue to the reality which is the truth of himself." [13] Bradley's ideal of self-realiza-

---

[12] Bosanquet, *op. cit.*, p. 131.
[13] *Ibid.*, p. 208. This is one of those typically Hegelian sentences from which, read it how you will, the precise meaning escapes.

tion suggests levels of development for the aspiring soul to pass through.[14]

Something of this sort does occur, and its occurrence is, I do not doubt, a major good. But, on Hegelian metaphysics, this can be only a shadow process: there is no real future, and the levels of development (or deterioration) through which you and I will seem to pass are already fully existent in the Absolute. The tedious fate of omniscience is that it can never be surprised.

(2) Suppose we say, next, that every evil occurring in space and time is also atoned for in space and time, so that the moral balance is always maintained and thus the "lesser reality" itself manifests the perfect justice of the Absolute. Royce, for example, held that the moral economy of the Absolute does not imply an entire absence of evil in the world of time and change. Rather, it implies certain acts of atonement which, paralleling every wrong deed, will keep the balance even. The atonement is not necessarily made by the doer of evil, but it must be made by somebody. In such a way, by such a balance, the universe displays its own perfection.[15]

This doctrine is really astonishing. In the first place, what do we want, atonement or correction? Atonement means that two people suffer instead of one. There may well be justice in this, but what we want is *remedy*. Punishment of Isaiah Nixon's murderers would have been atonement, but that leaves Mrs. Nixon far short of positive good. Evidently, evil requires not atonement but extirpation. A nice balance of sin and suffering

[14] *Ethical Studies*, New York, G. Stechert and Sons, pp. 59–60.
[15] See *op. cit.*, Vol. II, p. 368.

will not enable the temporal world to reveal the perfection of the Absolute.

Secondly, one may wonder about the ethics of a universe in which the innocent help atone for the deeds of the guilty. In working out this notion, Royce exhibits an almost masochistic zeal. He prefers not to know whose sin he may be atoning for, because such acts, regarded in the light of natural misfortunes, give opportunity for sacrifice and devotion. But if he cannot avoid knowing the sinner, he can then, with full awareness and rejoicing, share the general effort of atonement. Atonement, it thus appears, is a duty laid upon innocence as well as guilt.[16]

It is rather appalling, this eagerness to share in guilt. Perhaps so blameless a life as Royce's needed borrowed sins. But when, like the Hegelians, you carry the universe upon your shoulders, the weight of the burden crushes even as the passion exalts. The evil that all men know in their agony pours over you, and you suffer yourself to be made vile that the universe may be clean.

The failure of these efforts confirms what was our original view, that ethics requires an assumption of the reality of change. If, therefore, you take ethics seriously, Hegelianism is not the metaphysics for you; and, very oddly, it was not the metaphysics for Hegelians, who were not less fond of moralizing than anyone else. They were well aware of the paradox, their adversaries having displayed it to them often and very cheerfully. Unwilling, however, to liquidate the Absolute (the metaphor seems enchanting), they discoursed upon such virtue as would suit a static universe.

[16] See *ibid.*, p. 391.

Now, in a static universe those virtues vanish which have to do with action. Courage, temperance, foresight, even wisdom are out of place. Justice you might make a case for, but even so its work is already done in the Absolute. It seems, after all, that there is only one possible virtue, loyalty, and Royce devotes a whole series of lectures to it. His account is, unintentionally, a *reductio ad absurdum*, but at least he makes of the virtue nothing servile or base. Leslie Stephen once observed, prophetically, that "loyalty is a word too often used to designate a sentiment worthy only of valets, advertising tradesmen, and writers of claptrap articles."[17] But Royce's concept suffers from altitude, not baseness.

Loyalty, he tells us, is the attachment of the self to some cause which it selects by its own free will as the object of a lifetime's devotion. Normally, one would expect the merit of the virtue to derive from the merit of the cause: one would think that loyalty among friends rates higher than loyalty among thieves. Strange to say, however, this is not the Roycean idea. "Causes" belong to the temporal (and less real) world; loyalty, to the eternal. The usual hierarchy of values is turned upside down, and actions are judged not by their consequence but by their quality.

Since it doesn't matter what cause one is loyal to, the virtue is left to feed upon itself. The ultimate ethical maxim prescribes loyalty to loyalty. I can relieve the tautology of this maxim by understanding it to mean, as Royce does, that my loyalty to my cause must not cor-

[17] *Samuel Johnson*, in the English Men of Letters Series, New York, Harper and Brothers, p. 102.

rupt or destroy other people's loyalty to their causes. Unfortunately, causes conflict; and it is difficult to see how, when loyalty to my cause obliges me to attack somebody else's cause, I can possibly obey this rule. For if my attack is successful, I have destroyed for someone else the object of his loyalty; and if my attack is unsuccessful, I have not done much for the object of my own.

What happens to ethics, when actions are judged without regard to their consequences, may be read, written large, in Royce's application of the concept to human struggle. In his view, the special mischief in social violence does not issue from the killing, maiming, or starving of men. It issues from the fact that losers in battle lose also the object of their loyalty. They lose the chance to be loyal, and the spirit itself of loyalty withers within them.[18]

What an inversion of values! Surely the trouble with death camps was, not that they destroyed loyalty, but that they destroyed people. Surely the trouble with colonialism is, not that it corrupts the loyalty of natives (or, at any rate, not that alone), but that it maims, injures, pauperizes, and kills *them*. The moral of moral theory is plainly this: when you neglect human needs, you have found a means of escaping ethics; when you omit to consider results, you have bade farewell to the moral life.

And so, in our day, the frozen universe is uninhabited. Eager explorers described the grandeur of the waste and its total unlikeness to our little, warm, and buzzing world. Year after year, the tales were told and admired. But they never drew the multitudes from home.

[18] See *The Philosophy of Loyalty*, New York, Macmillan, 1916, p. 116.

Politics

As a mingling of social fact and legal fiction, political science is the most remarkable study invented by man. It gives an account of governmental practices and institutions; it also expounds a highly fanciful concept called "sovereignty." In England it proves that a cabinet whose policies are in the interest of monopoly ought, when sufficiently challenged, to be replaced by a cabinet whose policies are also in the interest of monopoly. In America it proves that monopoly ought to replace its representatives at stated intervals.

Sociologists, a less stable race, are chaotically aware of change. To them, as to infants, the world is full of lights and noises. But your political scientist, who was born old, has escaped the confusions of infancy and the heats of youth. He looks upon history, not as the nurse of new societies, but as a governess instructing mankind in the permanent political facts. Thus Plato's ideal of scientific administration by philosopher-kings is discovered to have been realized in a parliament of leisured gentlemen, and Aristotle's political animal fulfills his destiny in the recurrent convulsions of presidential campaigns.

Political science draws its peculiar character from the double meaning it gives to the verb "exist." When it describes, let us say, the behavior of an existing government, it is dealing with actual events in the world of space and time; but when it asserts that sovereignty "exists" in a monarch, or in a parliament, or in the people, it is talking about another kind of existence altogether. It is talking about legal existence.

Now, legal existence is what law or the theory of law asserts to be the case. And what is thus asserted to be the case is not necessarily what actually goes on in a given society. We are said, for example, to be born with natural rights, but no obstetrician would see them. The baby, grown up, might never be permitted to use them. Nevertheless, the rights would "exist" in him: so says our jurisprudence.

Again: Mr. Truman maintained that in the Presidency there "are" certain inherent powers. You could have searched him without finding them, because they are not observable objects in space and time. Their existence is legal existence, and it disappeared from the Presidency when the Supreme Court (or rather six of the members thereof) found it somewhere else. And they, in their turn, didn't find it as a scientist finds things; they got it by inference from constitutional theory.

Furthermore, legal existence, as asserted by jurisprudence, is permanent existence. Particular systems of law doubtless change by growth and decay, but what jurisprudence asserts to be the case *is* the case once and for all. In this manner, a certain set of concepts wears the aspect of eternity. Skeptics may well be right in thinking that these concepts are projections of the society which made them. Nevertheless, after the thing is done, society appears to be a projection of the concepts.

This sort of thinking is congenial to philosophical idealists, because it represents the triumph of thought over things.[19] For Hegelians it had the added charm

---

[19] Compare Bosanquet, *Social and International Ideals*, London, Macmillan, 1917, p. 85: "The magic of idealism lies, I suppose, in its

of asserting a permanent reality which is much more authoritative than the world of mere events. It suggests, that is to say, an eternal order of truth and righteousness such as Hegelians believe to be alone fully real.

One need not be cynical in order to anticipate what the applications of such a doctrine will be. An eternal order of truth and righteousness is plainly just the ground on which to base existing property relations, if these are what you want to defend; you can give them in this way a moral and legal justification, no matter how iniquitous they may otherwise appear. By thus identifying them with what is righteous *and* permanent, you make the opposition seem to be waging a struggle which is both wrong and impossible of success. Here is a recent example of the device:

> Most Americans accept Idealism as their creed. Likewise they reject Collectivism, believing it to be undesirable for them. Perhaps there is a connection between Pragmatism and Collectivism on the one hand and between Idealism and Individualism on the other. It seems we have two sets of twin ideas, not always identical and sometimes strange acting ones. But there is a discernible basic pattern of affinity between the ideas in each set. The outstanding collectivists denounce Idealism as an arch enemy of the people.
>
> To the Pragmatist natural law — in the sense that men are endowed by their Creator with certain inalienable rights as Jefferson wrote — is humbug, a favorite word of Mr. Justice Holmes.
>
> Having deprived men of all natural rights and relieved them of all natural obligations to one another it became philosophically easy to justify "might makes right" which

---

promise of victory for the human mind. Somehow mind is to triumph; to subdue the 'real' or the 'actual.' "

the Pragmatists do — and that is but a step to tyranny. Generally, that describes one of the sets of twins.

Now for the second set. To the Idealist those sacred words in the Declaration of Independence and those which refer to nature and nature's God, express true law just as Cicero expressed it twenty-five centuries [sic] ago. "True law is right reason conformable to nature, universal, unchangeable . . . It is not one thing in Rome and another thing in Athens, one thing today and another tomorrow; but in all times and nations this universal law must forever reign, eternal and imperishable . . . God himself is its author, its promulgator, its enforcer . . ."
Americans accept this idea.[20]

Do they? Perhaps. As we shall see in the next chapter, it would be hard to imagine a philosophy more "American" than pragmatism, a philosophy so highly national that it scarcely caught on anywhere else. Also, it would be hard to imagine men more resolutely anticollectivist than the pragmatists are, for some of them clearly regard collectivism as a subject for police action. Nevertheless, I think Mr. Koegel is right in finding the idealist, quasi-Hegelian strain in America too.

The question is, where does one find it? One finds it among chief counsel and others whose task it is to present American capitalism as immortal though embattled. One finds it also, and in massive form, among the multitudes who, exactly reversing Mr. Koegel's position, retreat from capitalism in the temporal world and cleave hopefully to some more perfect justice in the eternal. Thus one and the same doctrine serves property owners as justification and the multitudes as balm. Can a ruling class desire more?

[20] Otto E. Koegel, Chief Counsel to the Twentieth-Century-Fox Film Corporation, in *Vital Speeches*, Vol. 16, p. 413. (April 15, 1950)

This is the secret of the political theory which Hegelians expounded and of its long dominance over Western thought. This is why nothing less than a profound social crisis, in which change became acutely observable, could demolish the Absolute and restore to philosophy the stubborn sense of time. In the writings of the Hegelians such motives are deeply buried beneath abstract argument, and were unknown, I think, even to themselves. We shall proceed to uncover one or two of these motives, not as seeking to rattle the skeletons of skeletons, but as seeking to show that political philosophies don't escape immediate practical effects simply by being transcendental.

Transcendental philosophies, indeed, are a kind of poetry, and should be read as such. Their language, which in fact describes the snug realities of our usual world, is as faithfully metaphorical as any roses significant of cheeks. This description very well suits the theory of the State as a transcendental object, that is to say, as any entity "over and above" the citizens and having a life and even a mind of its own.[21] The Hegelians mean to say that the nation-state is an ultimate social reality, that it has absolute legal power over the citizens, but that this power cannot be legally used to thwart capitalist enterprise. *That* they say this is fact. *How* they say it is poetry.

It is a general principle with the Hegelians that the more inclusive an entity is, the more real it is, the more righteous and the more nearly perfect. From this it

---

[21] See, for example, Bradley, *Ethical Studies,* p. 167: the State is "the objective mind which is subjective and self-conscious in its citizens: it feels and knows itself in the heart of each."

follows that the State, being inclusive of the citizens, is more real than they and more righteous. From this, in turn, follows the State's legal power over the lives and fortunes of its citizens. And so Bosanquet wrote, "By the State we mean Society as a unit, recognized as rightly exercising control over its members through absolute physical power." [22]

This means that the State can legally require of the citizens any conduct it chooses; and if, in its mind and heart, it decided to turn Marxist, it could legally abolish overnight the institution of private property in the means of production. This revolutionary implication reproduces the historical fact that Marx himself sprang out of Hegel. The more orthodox disciples, however, retreated with some fright, and undertook to tame the wildness of their ruling principle.

Accordingly, Bosanquet tells us that the State, in its political behavior, is limited to "hindering a hindrance":

> The State is in its right when it forcibly hinders hindrance to the best life or common good . . . It may try to hinder illiteracy and intemperance by compelling education and by municipalizing the liquor traffic. Why not, it will be asked, hinder also unemployment by universal employment, overcrowding by universal house-building, and immorality by punishing immoral and rewarding moral actions? [23]

Why not, indeed? Unemployment and overcrowding are certainly hindrances to "the best life or common

---

[22] *The Philosophical Theory of the State*, London, Macmillan, 1910, p. 185. I think Bosanquet has one adjective in the wrong place: it is the right of control which is absolute, not the physical power.
[23] *Ibid.*, p. 192.

good," and there is nothing *in the theory* to prevent the State's dealing with them as with illiteracy and intemperance. The real reasons, though obvious, have nothing to do with the theory itself. Universal employment is a burden to employers; and universal housing, especially under state control, is to real estate operators a source of madness. Moreover, if it should turn out that private profit is a hindrance to "the best life or common good," Bosanquet's principle will imperil the whole capitalist system.

Evidently, limiting the State to "hindering a hindrance" doesn't limit it enough. We shall have to do better. Accordingly, Bosanquet goes on to argue that the State cannot do what is self-contradictory. If it can be shown that a policy of universal employment and universal housing is self-contradictory, then the State will not be able to undertake such a policy.

There is plainly no impediment of *logic* which can prevent the State from building homes and employing citizens; that is to say, the action is not self-defeating, nor the notion of it self-contradictory. Bosanquet is driven to say that the contradiction lies in its being "a direct promotion of the common good by force." This the argument holds to be impossible in the sense in which it is impossible to make people honest by legislating that they shall be so. I have to confess that the analogy escapes me, and that, on the contrary, "promotion of the common good by force" seems to me exactly what actual states assert they are engaged in. At any rate, it is quite clear that Bosanquet stops, though his principles bid him proceed, the moment his

PERMANENCE WITHOUT CHANGE 73

argument suggests a basic change in existing society. It is a prime example of the accommodation between theory and social pressure which we discussed in Chapter 2.

The same sudden halt for the same social reasons appears, much magnified, in Bosanquet's discussion of the nation-state. Historically, of course, the nation-state is a product of commercial society, and is, indeed, one of its most characteristic products. There have been many such states, and the competition among them has been deadly. But the only kind of world-state which this society realistically considers is that of the world dominated by one triumphant nation. In short, the nation-state enlarged into an empire is the widest horizon a capitalist ever sees.

Hegelian political theory stops at this point also. The nation-state, says Bosanquet,

is the widest organisation which has the common experience necessary to found a common life. That is why it is recognized as absolute in power over the individual, and as his representative and champion in the affairs of the world outside.[24]

And later on:

The object of our ethical ideal of humanity is not really mankind as a single community. Putting aside the impossibilities arising from succession in time, we see that no such identical experience can be presupposed in all mankind as is necessary to effective membership of a common society and exercise of a general will.

Well, then, the nation-state, and not humanity, is

[24] *Ibid.*, p. 320. The next quotation is from p. 329.

what must serve us as at once the goal and guardian of our destinies. This unexpected satisfaction with a limited concept suits very oddly with Absolute Idealism. For, as we remember, on this theory the more finite is less real, and the less finite is more real. This principle, regarded as axiomatic by Hegelians, would mean that, just as the State is more real than the individual (being less finite), so the State is less real than mankind. The conclusion ought then to be, not only that a world-state exists, but that it always has existed. And since loyalty to the larger body is obligatory, it must follow that our primary loyalties are to mankind. But if we took this doctrine seriously, we should have to be more concerned about the multitudes of men in Asia and Africa than we are about the profits of our own overlords.

Few things are more painful than the rigorous deduction of conclusions from postulates. Where is the man who can face such results? The Hegelians certainly could not. They were fast-bound to the society they lived in. When you showed defects in it, they came to its defense with metaphysics. When you showed the consequences of the metaphysics, they appealed to fact. When you showed the facts to be damaging, they appealed again to metaphysics. And so on, around and around.

Hegelians cannot be captured by linear chase. They have to be encircled. Yet I can imagine that, after the capture, when one leads them forth bound with chains of roses (for they deserve no worse), they will keep saying, with their old unembarrassed grandeur, "This is only your finite point of view."

Bernard Bosanquet

It is obvious from their political theory that the English and American Hegelians were quite content with the existing social order. Some, perhaps most, of them were reformers, yet with that skepticism of reform which their metaphysics inevitably implied. Bradley, for his part, was a high Tory, whose political zeal flamed hottest in a hatred of Mr. Gladstone. The fact that this hatred had been aroused by Gladstone's failure to rescue Gordon in the Sudan shows how thoroughly Bradley approved the ostensible aims, at any rate, of British imperialism. And when the Grand Old Man had grown very old indeed, Bradley observed with undiminished acerbity that "the devil was too much afraid of him ever to come for him." [25]

The most interesting of the group, however, is Bernard Bosanquet, because he wrote in detail upon specific social problems and participated in various movements of reform. He therefore serves as an accurate measure of how far an Hegelian might go in fostering that lesser reality, change.

Bosanquet's career as a reformer began quite early. He joined the Charity Organization Society in 1881, and his close association with it was crowned by his marriage, in 1895, to one of its leading members, Helen Dendy, who had already demonstrated the laudable concord between philosophy and social work by translating Sigwart's *Logic*.

The Charity Organization Society was founded upon

---

[25] Quoted in A. E. Taylor's memorial on Bradley, *Mind,* Vol. XXXIV, p. 6.

benevolence mingled with fear. There was a wish to help distressed paupers; there was also a fear, as Helen Bosanquet put it, of "indiscriminate charity sapping independence." [26] A reluctance to work, when it is workers who are reluctant, has always seemed immoral to capitalists. For an idle worker is a source of expense, but a working worker is a source of profit.

Moreover, in 1869, when the Society was founded, individualist theory was at its most fantastic extreme, and remained there for many years. In 1884, Herbert Spencer could still deplore (as *he* would in any year) the human sympathy which says of a pauper "Poor fellow!" when it ought to say of him "Bad fellow!" [27] In such an atmosphere the Society was constrained to deliver charity with due decorum.

As the Society developed, its views grew more rational, and the nonsensical moralizing disappeared. Bosanquet's views changed similarly. In a paper read before the Fabian Society in 1890, he had announced, "I want all ordinary cases of destitution to be treated in the workhouse, with gentleness and human care, but under strict regulation and not on a high scale of comfort." [28] Three years later, he had abandoned the distinction between deserving and undeserving paupers in favor of a simpler and more scientific classification according to

[26] Quoted in Helen Lynd, *England in the Eighteen-Eighties*, New York, Oxford University Press, 1945, p. 80.
[27] In an essay in the *Contemporary Review*. Helen Lynd quotes the passage, *op. cit.*, p. 91.
[28] "The Antithesis between Individualism and Socialism Philosophically Considered" (a title, that!), published in a collection of his papers called *The Civilization of Christendom*, London, Swann Sonnenschein, 1893, p. 342.

"age, sex, behavior, physical condition, and sensitiveness." [29]

During the next twenty-five years, a mild Laborite tone steals into Bosanquet's writings — not, indeed, into what may be called his official works, but into his papers and addresses. He now speaks of the workers as a social class and of their class-loyalty as an admirable trait.[30] He is able to contemplate socialist theory without anger, and he recognizes some of the attacks upon it to be "not fundamental criticism but malicious hypothesis." [31]

He is aware, also, of the effect of imperialism upon native populations. I do not know that he actively opposed it, but he is careful to say that this is not what he means by patriotism, and he is bold enough to publish this thesis in the midst of an imperialist war. He understands, and he accepts, the rueful dictum of the African chief: "First the missionary, then the trader, then the gunboat, and then — oh Lord!" [32]

Such a man, however, is not yet a revolutionary. For that a further step remains: to perceive that the liberation of the workers and of the colonial peoples is one and the same historical operation, requiring something more than sympathetic approval of the two groups. This step Bosanquet did not and, I suppose, could not take. The providence which established his birth in the year

---

[29] *Ibid.*, p. v.
[30] As, for example, in *Social and International Ideals*, London, Macmillan, 1917, p. 193: "It is not a career for himself, but recognition, security, status, and power for his class that the loyal and class-conscious working man desires."
[31] *Ibid.*, p. 216.
[32] *Ibid.*, p. 16.

1848 allowed itself the gentle irony of making him an Hegelian, not a Marxist.

Instead, Bosanquet came to look upon the workers as a group whose struggles he admired and whose sufferings he helped allay. Since his vision, which was otherwise humane and comprehensive, hardly embraced a socialist world, he was left in magnanimous contemplation of sheer conflict ennobled by the presence of ideals. And so he wrote in a review of Sorel's *Reflections on Violence:*

> Give us, we are inclined to cry, in every class or functioning organ of the community, such a faith and inspiration as he [Sorel] claims for the workers and their gospel, and we could have confidence in the future, not because we could predict the detail of what must come, but because whatever comes, under the influence of such inspiration, and to a people so prepared to suffer and be strong, could not be other than good.[33]

To suffer and be strong! In our present years the words ring out less nobly than when Bosanquet wrote them. They came to express, as he certainly could not anticipate, a favorite doctrine of fascism, which taught the masses to value suffering and to forget that the gains of conflict were to belong to someone else. It is a morbid and unnatural idea. Doubtless the hope that suffering will be a source of strength is the chief comfort of those who suffer. But the truth is that the sufferers will grow strong in proportion as they cause their sufferings to cease.

Thus Bosanquet's social views show us how far Hegelianism can go without ceasing to be Hegelian.

[33] *Ibid.*, p. 188.

PERMANENCE WITHOUT CHANGE 79

The theory, we discover, can be aware of specific injustices and can provide its believers with a modest rationale of improvement. It can produce reformers who know the limits of reform and who, within those limits, live content. But it envisages no vast social reconstruction. It looks upon the strife of classes, not as an historical movement toward a new society, but as an occasion for the display of fortitude and loyalty. And with most of the Hegelians this struggle is deplored,[34] or unmentioned, or altogether unseen.

Was it Bosanquet's Hegelianism that kept him from going further, or a wish to go no further that kept him in Hegelianism? I cannot tell: the accommodation of the actual philosophy to the possible wish is far too perfect. But at least one knows that there was nothing in Hegelianism to compel advance and much to suggest the greater comfort of standing still.

With the "right" philosophy, one can gather patience while the world is sad. "It is certain, to my mind," wrote Bosanquet, "that evil and suffering must be permanent in the world, because man is a self-contradictory being in an environment to which he can never be adapted, seeing that at least his own activity is always transforming it."[35]

How narrowly truth is sometimes missed! For the human activity which transforms the world can make it a place in which evil and suffering are so negligible as to render the question of their permanence merely

---

[34] Royce, for example, did so when he said that loyalty, properly understood, does not cause enmity between social classes. See *The Philosophy of Loyalty*, p. 214.
[35] *Social and International Ideals*, p. 300.

academic. But unless this truth is understood, the human activity which transforms the world will be largely that of exploiters.

First the philosopher, then the politico, then the commentator, and then — oh Lord!

CHAPTER IV CHANGE WITHOUT
            PERMANENCE

SPECULATION, when it is of the mind, is a noble activity. It is also tiring. Few men and few epochs have been able to sustain that amplitude of vision which looks throughout the universe and sees it all at once, whether as a changeless system with multitudinous parts, or as a movement of events with discoverable laws. Knowledge *sub specie aeternitatis,* as Spinoza loved to say, is highest knowledge. But at that height, forest and grassland have fallen away, the air is thin and cold, and even the hardiest spirits must visit briefly.

Those visits, however, have been of infinite value to mankind. Without them, there would be no theory of relativity, no quantum mechanics, no evolutionary hypothesis, no rational conception of history. Without them there might be a knowledge of many things, but no understanding. Even when the visitors proved wrong, they nevertheless demonstrated that the climb could be made, that the altitude could be suffered, and that the view was of something more than mist.

Meanwhile, the more earth-bound mortals conduct their ceaseless investigation of detail. They "number the streaks of the tulip or describe the different shades in the verdure of the forest." Their labor too is often of value, for it corrects inaccuracies.

Vagueness is the peril of large vision; narrowness is the peril of small. Only the greatest minds have been able to unite the two visions so successfully as to mitigate the faults of each. Other minds, doomed by nature or by social station to larger degrees of error, are left to work out their destiny as best they can. This working out, indeed, they find not wholly uncongenial; and, as habit confirms the practice of a lifetime, they end by producing philosophical justifications of all they have done.

When, at the opening of the twentieth century, Hegelian voices began to seem monotonous and the Hegelian vision began to lose its charm, there were murmurs from the meadows and shriller plaints from the forest. At first a mere cacophony, the sounds assembled themselves at last into recognizable structures. In the fullness of their special perfections, they chanted antiphonally, and sometimes howled harmoniously, a dirge for the dying Absolute.

The murmurs came from a species of philosophers called realists, who, in their modest prosperity, seldom spoke much above a whisper. The shriller plaints came from the pragmatists, who, even in adversity, seldom spoke much below a roar. In America they have kept their eminence to the present day by the simple, and yet extraordinary, faculty of making themselves heard. Nevertheless, because articulation is likely to be obscured by volume, there remains some doubt what the pragmatists actually say. And the doubt is deepened by the fact that whenever one arrives at a precise

CHANGE WITHOUT PERMANENCE

formulation of pragmatism, the pragmatists are accustomed to observe that they never said anything of the kind.

In the seventh decade of its life, pragmatism has acquired a kind of youthful venerability. Its protagonists, indeed, display upon their splendid bosoms as many arrows as Saint Sebastian, but without a comparable martyrdom and probably without hope of canonization. The wounds have proved somehow not mortal, and Sebastian may be heard explaining to his adversaries the uselessness of further shots.

The whole manner of resisting criticism is in fact remarkable. Pragmatists not only will disavow the statements you criticize but will embrace as their own the very point of view from which you criticize them.[1] Thus the critic finds himself not so much defeated as absorbed. He can avoid this fate, I should suppose, only by making himself an extremely nauseous morsel. That would be a novel way in which to philosophize, but it would have, appropriately enough, a pragmatic justification.

The movement of thought which at first called itself "pragmaticism," and thereafter gracefully dropped the penultimate syllable, came into being as a revolt. "Damn the Absolute!" cried William James to Josiah Royce as they sat on a stone fence, being photographed by James's daughter. The Hegelian armor, impenetrable by argument, could not even be scratched by mirthful profanity. But there was more than mirth in James's utterance, as

---

[1] Dewey did as much with Santayana and Russell. See *The Philosophy of John Dewey* edited by P. A. Schilpp, Chicago, Northwestern University, 1939, pp. 530–532 and 544–549.

Royce, after years of friendly controversy, must have known very well.

The Hegelians had seemed to place reason upon the side of a static universe. Now, the concept of a static universe was precisely what James hoped to demolish. In his earlier attacks upon it, he conceded it the use of its own weapons, and proposed for his own part to use weapons of a different sort. Where the Hegelians had used reason, he would use faith. It was like attacking an artillery emplacement with slingshots.

When experience confirmed his suspicion that the Hegelians had not got a monopoly of reason, James skillfully seized that instrument for himself. Meanwhile, he had fastened upon pragmatism its characteristic doctrine, that the truth of a sentence is to be estimated from the effects produced by one's believing it to be true. James had called this "the will to believe"; but, in view of its immediate philosophical motives, it seems to have been rather the will to disbelieve — to disbelieve, that is to say, in a universe where man's best labors can effect nothing.[2] To this potent negation we partly owe the fact that, among our manifold anxieties, we need no longer include the fear of a frozen world.

The Tricksy Spirit

William James is the only American philosopher — perhaps the only philosopher — to whom one can apply the adjective "adorable." No thinker that I know of

---

[2] James's phrase for this was "a negative will-to-believe." *The Letters of William James*, Boston, The Atlantic Monthly Press, 1920, Vol. II, p. 269.

has contrived to be so genuinely and universally affectionate, so eager to find good in people, and so readily persuaded that he had found it. His latter years were filled with lame ducks and lost causes, and on behalf of both he risked his purse and his reputation.

James never allowed himself that exemption from social responsibility which is claimed by superior intellects when they are also inferior hearts. Once, when he was seeking help for a needy, seedy metaphysician, he wrote, "Most men say of such a case, 'Is the man deserving?' Whereas the real point is, 'Does he need us?' " [3] One such philosopher will atone for a dozen Herbert Spencers. If pragmatism had been sustained by comparable spirits, its ethics would have been sublime.

Of all James's creations perhaps the most wonderful was that superb expository instrument, his prose style. Lithe and vivacious, it had that gift the lack of which Dr. Johnson lamented in the metaphysical poets: the gift of metaphors at once improbable and apt. No one but James, on seeing an enormous, fuzzy, and affectionate dog, could think of writing, "He makes on me the impression of an angel hid in a cloud. He longs to do good." [4] No one but James would be likely to compare the Hegelian universe to a seaside hotel, where no one has any privacy.

Upon all his writings James bestowed so intimate a relish that even in the more abstruse passages one can feel the man moving within the argument. This relish, I suppose, is most to be found in his letters, where he recorded with infinite wit everything that he found in

[3] *Ibid.*, Vol. II, p. 178.
[4] *Ibid.*, Vol. II, p. 26.

the American scene. For example, in July, 1896, he delivered at the Chautauqua Assembly a series of lectures later published as *Talks To Teachers*. His impressions of Chautauqua are set down in letters to his wife; and anyone who is, as I am, old enough to remember the fearful solemnity of that institution will share James's feelings. Here are a few passages:

July 24. I've been meeting minds so earnest and helpless that it takes them half an hour to get from one idea to its immediately adjacent next neighbor, and that with infinite creaking and groaning. And when they've got to the next idea, they lie down on it with their whole weight, like a cow on a doormat, so that you can get neither in nor out with them.[5]

July 27. I took a lesson in roasting, in Delsarte, and I made with my own fair hands a beautiful loaf of graham bread with some rolls, long, flute-like, and delicious. I would have sent them to you by express, only it seemed unnecessary, since I can keep the family in bread easily after my return home.

August 2. I have seen more women and less beauty, heard more voices and less sweetness, perceived more earnestness and less triumph than I ever supposed possible.

It is marvelous, this succession of pages — essays, books, letters — which carry the reader on wings over a landscape rich with ideas. The great trouble with the Hegelians was that one always knew what they were going to say. But one never knew what James would say, except that it would be fresh and original and usually penetrating. The ordinary foes of system-building set up a formidable cackle, as of hens denouncing the coop.

---

[5] *Ibid.*, Vol. II, p. 41. The other quotations are from the pages immediately following.

But James was an eagle that really did need space to fly. Politically, James was a being no less rare: he was a genuine liberal. That is to say, he embraced the authentic values of middle-class society, was aware that those values were frustrated, but did not realize that middle-class society itself frustrated them. In his letters he attacks President Cleveland's bellicose utterances in the dispute with England over Venezuela, he denounces American intervention in the Philippines, and he defends Dreyfus. But he seems to have had no notion of the commercial rivalries which underlie international conflicts and race prejudice.

The Paris Commune of 1871, which showed that there could be such a thing as a workers' government, drew from James no profounder observation than that "the gallant Gauls are shooting each other again."[6] And the Haymarket riots appeared to him to be "the work of a lot of pathological Germans and Poles."[7] Thus, though his sympathies were prevailingly with what are called the "legitimate aspirations" of labor, he was never very far left of center.

But his liberalism was free of humbug. An invitation to meet Kipling, which he could not accept, brought forth a definitive commentary on that poet's imperialist ideas: "If the Anglo-Saxon race would drop its sniveling cant it would have a good deal less of a 'burden' to carry. We are the most loathsomely canting crew that God ever made."[8]

[6] *Ibid.*, Vol. I, p. 161. James was at this time twenty-nine years old.
[7] *Ibid.*, Vol. I, p. 252.
[8] *Ibid.*, Vol. II, p. 88.

In this spirit James sided with the Boers in their struggle against the British. He chanced, indeed, to be in England in 1900, when proposals were being made for a day of national humiliation and prayer. Arms having proved ineffectual, the God of Battles was to be invoked on behalf of empire. Says James:

> I wrote to the "Times" to suggest, in my character of traveling American, that both sides to the controversy might be satisfied by a service arranged on principles suggested by the anecdote of the Montana settler who met a grizzly so formidable that he fell on his knees, saying, "O Lord, I hain't never asked ye for help, and ain't agoin' to ask ye for none now. But for pity's sake, O Lord, don't help the bear." [9]

The *Times* ignored James's letter, and thus preserved the freedom of the press for Western civilization.

A man who undertakes the profession of philosophy has many ideals to imitate. He needs Plato's breadth and Hume's clarity, Spinoza's noble self-reliance and Kant's subtlety of detail. These are tremendous acquisitions, and he will probably not get them all. But, however he fares with these, he will need James's humanity, if he is to overcome his own narrowness by an interest in other people's ideas, and if he is to treat philosophy as a thing of importance to human life. When he has achieved all this, he will be entitled to say, with James, "As for myself, I have ceased to be a humbug. I stopped teaching last June." [10]

---

[9] *Ibid.*, Vol. II, p. 118.
[10] In a letter to Professor Joseph A. Leighton, who has been gracious enough to send me word of it.

The Universe with the Lid Off

The motives of James's philosophy, as he himself understood them, were mainly ethical. That is to say, he sought a universe in which ideals could be concretely realized by human effort, and he viewed with invincible distaste the Hegelian notion of a universe already perfect. Evil is not to be got rid of by "subsumption" under a higher category of the Absolute, but "by dropping it out altogether, throwing it overboard and getting beyond it, helping to make a universe that shall forget its very place and name." [11]

Behind the ethical motives stand the political and social motives which we have just surveyed. There is in James an antipathy to "bigness and greatness in all their forms," [12] a somewhat Rousseauistic belief that "*every* great institution is perforce a means of corruption — whatever good it may also do." These are doctrines of philosophical anarchism, and James's pluralistic universe is just the metaphysics which such a social theory requires.

There is, of course, some truth in the theory. Institutions *do* corrupt, if only because they induce that sort of official lying which is called "public relations." Nevertheless there can be no social life without institutions nor any social change without mass action, which involves some limiting of private wills.

Consequently, philosophical anarchism tends to dull

---

[11] *Pragmatism*, New York, Longmans Green, 1907, p. 297. The sentence is in italics in the original.
[12] *Letters*, Vol. II, p. 90. The next quotation (in which the italics are James's) is from the same volume, p. 101.

the hope of actual progress by making an exchange of institutions seem like an exchange of evils and by interpreting every restraint upon personal freedom as a sign that freedom is about to vanish altogether. These pessimistic implications, however, were submerged beneath James's natural buoyancy and were left to appear in the writings of later pragmatists as a means of berating Soviet socialism.

If philosophical anarchism, as social theory, is a revolt against institutions, its metaphysics is a revolt against *system*. James felt about the Absolute the way an anarchist feels about the government: away with it, at once and forever! In the mid-twentieth century one can tolerate the Absolute as one tolerates a relic; but in 1900 the Absolute was a living Goliath, and its adversaries were necessarily embattled. However, the intensity of the struggle somewhat darkened strategy. Fifty years before James, Marx had turned Hegel right side up and quietly walked off with him. But James did not want companionship; he wanted atomization.

If Hegelians will forgive me a comparison which is perhaps uncomfortably close, I would say that James regarded Hegel in the way that Western thinkers now regard Lenin, as a dead and errant genius. But James regarded Bradley as those same thinkers regard Stalin, as a talent dangerous and alive. With Bradley at hand, etiquette might constrict but could not cool James's polemical ardor. There exists one essay containing a long quotation from Bradley punctured, as with bullets, by Jamesian rebuttals in brackets and italics, and containing also, towards its end, the scarcely apologetic observation, "Polemical writing like this is odious." [13] Odious, perhaps; but how else could one stanch the flow of Hegelian talk long enough to interpose a word? To James, the Absolute was all blemish and no charm. It was stuffy, it was dull, and it was everywhere. It was a sort of cosmic Chautauqua, blameless but somehow not to be admired. It was also, at least in part, a rubbish heap. Since it contained everything, it contained much worthless lore. "The rubbish in its mind," wrote James, "would thus appear easily to outweigh in amount the more desirable material. One would expect it fairly to burst with such an obesity, plethora, and superfoetation of useless information." [14] Where Bradley had watched the seasons simultaneously prevail, James perceived the humble, salutary movements of trash-collectors.

Now, it seems that if governments were suddenly abolished, the repressed individualities of men would leap and gambol beneath the sun. And it seems, likewise, that if you suddenly obliterate a system, the parts will roll or dart or swim in happy liberation through a friendly void. Neither of these results is in fact probable, and their seeming so is due to the supreme illusion of modern thought, that a system and its members have no essential connection with each other. Illusion though it is, this concept has fathered many philosophies, and James's pluralism is one of them. Let us watch this logic at work.

If James could ponder Mrs. Nixon's troubles — and

[13] "The Thing and Its Relations," in *Essays in Radical Empiricism*, New York, Longmans Green, 1912, pp. 111–116. The quotation is from p. 121. Copyright 1912, by Henry James, Jr.
[14] *A Pluralistic Universe*, New York, Longmans Green, 1916, p. 128. Copyright 1909, by William James.

we may be sure he would ponder them sympathetically — his first act would be to warn her against Royce and Bradley. Where there is no change there can be no improvement; to this critique we have already assented. What Mrs. Nixon needs is a world full of possibilities, among them the possibility of happiness for herself. Such a world must not be *fated;* the events occurring in it must not totally repulse human control. She must know that there is a chance and that she can seize it.

To this doctrine also we may assent. The problem is, how shall this sort of universe be conceived? James thought that if you relaxed the system, possibilities would come flooding in. The soil then might drink and be fertile, and the inhabitants, feeding on fruits instead of dust, might be for the first time fully men. However, James, who preferred commercial to horticultural metaphors, put it thus:

> That the universe may actually be a sort of joint-stock company of this sort, in which the sharers have both limited liabilities and limited powers, is of course a simple and conceivable notion.[15]

Conceivable, I would say, but not simple. There are varying degrees in which the system may be relaxed, with varying consequences answerable to the degrees. For example, if the relaxation were total and chance became a radical fact, this would introduce into the universe a blind, unpredictable, irrational force which no planning could possibly master. The world would then behave with singular caprice and contrary to all our knowledge of it.

[15] From "The Dilemma of Determinism," in *The Will to Believe, and Other Essays,* New York, Longmans Green, 1899, p. 154.

One might suppose that this was what James meant by his famous phrase, "a universe with the lid off." But a survey of his entire writings does not support so extreme a view. In the cosmic joint-stock company — Universe, Inc. — the members are bound by certain rules, which, however, arouse their private initiative and make profits thrive. You might say that in Bradley's universe there was one eternal balance sheet, but in James's you could watch the weekly curve go up and down.

The curious thing about James's philosophizing is that his destinations are clear and his arrivals are uncertain. I have read him a good deal and have a lively impression of what he wanted to say, but I have no such impression of what he actually said. I believe that a listing of his statements on the nature of change would exhibit every conception from order to chaos. His thinking, as Santayana very aptly said, was improvisational. That wonderfully sympathetic mind understood every point of view except its own.

Perhaps we must resolve the problem in terms of emphasis. We can say then that, in James's view, the universe is fluid, with multitudinous possibilities, so that at any rate the *deepest* cravings of men have ample room for fulfillment. I think that James took these cravings heterogeneously rather than systematically. That is to say, he wanted every deep craving satisfied, no matter how contradictory they might be in their total mass. This wish is what forces upon his metaphysics the appearance of a belief in pure chance, since only an irruption of that sort would suffice to suspend the regularity of nature and the law of contradiction. Thus the

Jamesian universe seems like change without permanence; it dissolves all solid footing, in order that men may soar.

But, as we have seen, no inference from any one passage in James's writings will prove tenable for all the others. We owe it, then, to him (and to our own enlightenment) to quote one long paragraph, which shows how well he could understand the universe when it was a question of correcting someone else's mistakes. The paragraph describes Hegelian dialectic, not (in the Hegelian manner) as a relation of categories, but as a relation of events in a world of change; and it is the best such account that I know of.

The impression that any *naïf* person gets who plants himself innocently in the flux of things is that things are off their balance. Whatever equilibriums our finite experiences attain to are but provisional. Martinique volcanoes shatter our wordsworthian equilibrium with nature. Accidents, either moral, mental, or physical, break up the slowly built-up equilibriums men reach in family life and in their civic and professional relations. Intellectual enigmas frustrate our scientific systems, and the ultimate cruelty of the universe upsets our religious attitudes and outlooks. Of no special system of good attained does the universe recognize the value as sacred. Down it tumbles, over it goes, to feed the ravenous appetite for destruction, of the larger system of history in which it stood for a moment as a landing-place and a stepping-stone. This dogging of everything by it negative, its fate, its undoing, this perpetual moving on to something future which shall supersede the present, this is the hegelian intuition of the essential provisionality, and consequent unreality, of everything empirical and finite. Take any concrete finite thing and try to hold it fast. You cannot, for so held, it proves not to be concrete at all, but an arbitrary extract or abstract which you have made from the remainder of empirical reality. The rest of things invades and overflows both it and you together, and defeats your rash attempt. Any partial view whatever of the world tears the part out of its relations, leaves out some truth concerning it, is untrue of it, falsifies it. The full truth about anything involves more than that thing. In the end nothing less than the whole of everything can be the truth of anything at all.

Taken so far, and taken in the rough, Hegel is not only harmless, but accurate.[16]

Expediency and Truth

"Nothing less than the whole of everything can be the truth of anything at all." For my part, I have no doubt that this is the case, but I think it little less than astounding that a pragmatist should say so. The whole point of pragmatism (if indeed pragmatism has a point) is that truth is what satisfies, not some universal condition of things, but some immediate need of your own. On this theory, truth is opportunistic: you make a quick clean-up, get out of the market, and present yourself as the greatest scientist of the age.

James's career as a pragmatist, in which of course lies his chief fame, thus prevented any effective use of the insight he had so shrewdly discerned in Hegel. The pragmatic theory properly goes with a metaphysics of chance, since in a world where anything can happen one of the things that can happen is that myths are true whenever they are useful. But this sort of venturesome nonsense has nothing to do with dialectics.

[16] *A Pluralistic Universe,* pp. 88–90. The italics are James's. The dropping of capitals on adjectives made from proper names is also James's, one of his experiments with revolution.

The theory of knowledge is known to philosophers as epistemology: "portentous name and small result," as James dryly observed.[17] We shall examine it in some detail in the next section of this book. We must, however, give some attention to it here, because the pragmatic version of it shows what happens when the universe is conceived in too loose and, one might almost say, too casual a style. It also shows that Mrs. Nixon can hope for nothing but good wishes from the pragmatists.

James's treatment of epistemology broke upon the world with an air of something very like levity. The speculations of previous philosophers had manifested an immense sobriety, such as the subject seemed to demand. But James assailed it with a battery of astounding metaphors, drawn from the stock exchange, the commodity market, and the race track — that is to say, from gambling in general. He now began to talk of "cash-value," of live and dead "options," of "backing the religious hypothesis against the field." [18]

Remembering James's trick of remote allusion, we shall be wise not to take these metaphors as a sign of commercial intent. Nevertheless, it is difficult not to feel, as many commentators have felt, that in American pragmatism we have a philosophy which frankly confessed its economic origins. For, in any other than a commercial culture, such metaphors would be not only remote but impossible.

[17] *Letters*, Vol. II, p. 121.
[18] The first of these metaphors will be found *passim* throughout *Pragmatism* (e.g., p. 53) and other works. The last two metaphors are from *The Will To Believe*, pp. 3 and 26 respectively.

When James talked about cash-value, he wanted you to feel epistemology in your pocketbook, a place of sensitive reflection. There *is* a connection between truth and human behavior: the fact would have been plain enough, except that the Hegelians, with their eyes on the clouds, and the empiricists, with their eyes on the pebbles, had tended to obscure it. And when James talked about backing hypotheses, he wanted you to know that what goes on in the universe is no more predetermined (or at any rate no more predictable) than what goes on in a horse race. The metaphor is not altogether grotesque, but it suits somewhat different conclusions. For scientists will bet only when the race is between Man-o'-War and some faltering nag, whereas pragmatists, in their zest for winning, are very likely to try fixing the race.

It is now too late to remedy the impression that pragmatists believe in a free and easy universe in which the inquirer may play fast and loose. Once the theory of knowledge was resolved *entirely* into questions of method, truth-getting became a matter, not of objective reality, but of subjective mechanics. You could make truth as you made commodities, and the raw material of it was genial enough to accept any shape you wished. The quite correct notion that human inquiry is set by human need degenerated first into the notion that inquiry is restricted to *immediate* needs, and then to the notion that whatever sentence "satisfies" a need is true.

Now, it is fairly notorious that needs can sometimes be satisfied by illusions; in fact, that is the precise psychological nature of insanity. The lunatic, unable to adapt to the world as it is, imagines the world to be of a sort he can adapt to; and thus he finds his harmony. But on the supposition that any belief which satisfies a need is true, the difference between sanity and insanity disappears. We shall have to add scientists to Shakespeare's list of the lunatic, the lover, and the poet, who are "of imagination all compact."

Pragmatists, of course, have always contended that they meant no such doctrine as this; and we could believe them, if only they had revised their account so as to eliminate this possibility. For evidently the correct account is, not that any belief which satisfies a need is true, but that needs are properly satisfied (if they can be at all) only by true beliefs. Consequently, the standard which determines whether a sentence is true is something other than the need which is to be satisfied.

It is, therefore, rather less than comforting to be told by James that " 'The true' . . . is only the expedient in the way of our thinking, just as 'the right' is only the expedient in the way of our behaving."[19] It would be easy to attack this sentence with moral lightnings, but we may be sure James meant nothing ignoble by it. Nevertheless, the very next clause tells us that "expedient" means "expedient in almost any fashion." The range is certainly very wide, and we are left wondering where the line is drawn. Apparently, "valuable" beliefs fall within this area, regardless of their factual ground, since James says, "If theological ideas prove to have a value for concrete life, they will be true, for pragmatism, in the sense of being good for so much."[20] Such language is invincibly vague; but if it means anything, it

[19] *Pragmatism*, p. 222. The sentence is in italics in the original.
[20] *Ibid.*, p. 73. James put this sentence also in italics.

means that a belief is true, provided it makes you happy.

The gay abandon of the pragmatic theory of knowledge is made more solemn but not less free in James's account of live and dead options. This account (a fable, as I think, set in a philosophy) runs somewhat as follows. When a choice between two propositions is important to us, we have what James calls a "live option." We cannot then decline to choose, as we might if the alternatives were trivial. It is *psychologically* possible, of course, to avoid choosing, but what we cannot escape (so James thinks) is the moral demand that a choice be made.

The classic example of such an option is the two sentences "There is a God" and "There is not a God." The alternatives, James thinks, are momentous in the sense that either one will have profound consequences for human life. At the same time, the question is one on which science has nothing definitive to say.

The issue being undecided, we can proceed by a kind of wager; and, like courageous bettors, we shall pick the horse which, having a chance to win, will bring the most reward. Of the two alternatives, the sentence "There is a God" is clearly the proper choice. It has a fifty-fifty chance of being true; and if it turns out to be true, it will bring the believer all the benefits which accrue to such believers. If it turns out to be false, the believer will be in no worse case than if he had chosen the other alternative. Moreover, he escapes the risks of believing the other alternative and having it prove false, the consequences of which are hell-fire.

This argument, which James borrowed from Pascal, seems to me conceived in desperation, and gives the unlovely picture of a man choosing his beliefs for the sake of personal safety. The argument has perhaps some plausibility when the alternatives are merely two; but the odds become dangerous at three, and fatal, I should think, at any larger number. Suppose we take these three assertions: "Mohammedanism is the only true religion," "Christianity is the only true religion," "Buddhism is the only true religion." Suppose, further, that we have no reason to think any one of these truer than another. Then, when we come to "bet," the odds are, not even, but two to one against any of the possibilities. To overcome this risk the rewards would have to be prodigious. The eight-foot heavenly houris that Mohammedanism offers certainly seem more enticing than Nirvana; but I doubt whether, apart from more persuasive argument, they would produce much betting.

The fact is that James had a real problem but no real solution. By the end of the nineteenth century it had become plain that many beliefs which claimed certainty did not in fact possess it. What they did sometimes possess, however, was probability — in the twilight of which (as Locke finely said) we live most of our lives. But, instead of developing probability theory, where the answers to all such questions lie, James embarked upon a romantic excursion which was to reveal a nice adaptation of truth to human needs. The purpose was extremely liberal and humane, since everyone's deepest convictions were somehow to be validated. But, in the process, truth and probability were so far unhinged that there remained no very good reason for believing anything.

It was a sad result, which later pragmatists have only

worsened. Yet, in the history of thought, bad answers are sometimes providential: they show how good the question was. They also show how the question should have been put. Accordingly, among our several debts to pragmatism there is this: that it demonstrated the fatal metaphysics (if I may use a word distasteful to pragmatists) of conceiving the world so freely as to make it positively licentious. The lidless universe, with all its gusts and sudden fires, is thrilling and spectacular, and might even be safe. But I very much doubt that it could ever be known.

CHAPTER V PERMANENCE
THROUGH CHANGE

THE TWO PREVIOUS CHAPTERS, taken together, propound this lesson: without change, no ethics; without order, no knowledge. In the Hegelians' view of the world there is order but not change; in the pragmatists' there is change but not order. Consequently, neither theory yields knowledge and ethics simultaneously, and neither succeeds in offering a rational means of improving man's place in nature. Mrs. Nixon might be comforted, but she would not be much helped.

This fact we have, as I hope, elicited from the two philosophies. Our treatment of each, however, was rather more expository than critical. We did not reach the point of demonstrating, as far as we might have, that the theories are erroneous. We have said merely that we didn't like some of their implications, and have given the reasons why.

Propagandistically, this might seem sufficient; that is to say, we have perhaps said enough to persuade people not to take these views too seriously. But, philosophically, it is not sufficient. We are required to show that the treatment of change in the Hegelians and the prag-

PERMANENCE THROUGH CHANGE

matists is false as well as frustrating, that (in Talleyrand's famous phrase) "it is worse than a crime; it is a mistake."

To do this, we must set up a contrary conception which has a greater probability of being true. As against the Hegelians, we have to show that change is at least as real as permanence, and that accordingly the universe is not a static, absolute, spiritual entity. As against the pragmatists, we have to show that the universe is by no means as indeterminate as they suppose it to be, and that on the contrary the very determinateness of things — their *rigor,* if you like — puts change under human control. This gives us two main divisions for the present chapter.

I propose to attack the Hegelians on a very broad front — a front so broad, indeed, that it will include some of their opponents also. The attack will fall upon the entire conception of "supernatural" entities. By way of definition, let us say that the term "natural" applies to things which move in space and time and are at least potentially observable by human sense organs. (The second clause permits us to regard as natural certain things, like remote stars, which have not yet been observed but may sometime be observed.) Then the term "supernatural" will apply to things not characterized by motion in space and time and not observable by human sense organs, even potentially.

Now, by these definitions the Absolute is, of course, a supernatural entity. It is said not to be in space and time, but to contain space and time in it as part of "the hazards and hardships of finite self-hood." Accordingly, if we can show with a high degree of probability that real existence is bounded by space and time, then the Absolute, falling outside the reach of these coordinates, falls also outside existence. Thus we would justify our previous assertion that the conception of the Absolute is not science but poetry.

Belief in the supernatural has, throughout history, had many supports, among which, since the time of the Egyptians, is the power of priestcraft. Tradition, legend, and fanciful conjecture in the absence of knowledge have had a good deal to do with it, too. But the modern age, though not altogether free from gross superstition, has required a much more sophisticated support for the doctrine. Philosophically at least, it has preferred to base the supernatural not on the authority of scriptures or of church councils, but upon the supposed fact that no object can exist unless it is in some manner experienced or known.

The simplest (and perhaps best) account of this is in Bishop Berkeley's celebrated doctrine that an object, in order to exist, must appear in my consciousness, or in yours, or in someone else's, or in God's. Hegel's own version was less naïve but more vulnerable: since an object is always identical with what you can say about it, and since what you can say about it is of various sorts (called "categories"), these categories, embraced by a single universal idea, can be thought of as constituting the universe. Whatever the version, the doctrine always is that "existing" and "being known" are synonymous terms.

This is the doctrine we shall try to refute by demonstrating its exact contradictory, namely, that there are at any rate *some* things in the universe of which it is true that their existing does not necessarily imply their being known.

Is There Such a Question?

To get at this problem we have to overleap three walls, which are used to fortify it against discussion. The first wall is erected by common sense. To the question, "Are there things which exist without being known?" common sense replies, "Of course." Now, I think that common sense is quite right in so answering. Our task, however, is to prove this, if we can; and we shall have to offer something better than the hunches and guesses which common sense ordinarily displays.

The second and third walls were erected by pragmatism and positivism respectively, and each of them bears a sign reading TRASH. The sign warns us that the question is not worth discussing, and that philosophers, if they cannot be deterred by common sense, ought to be deterred by sophistication.

Pragmatism says that since you can't use a thing until you know it (or are, at any rate, aware of it), you are simply wasting time if you wonder whether it exists when you don't know it. To this we may reply that, although you must know what iron is and where it is before you can use it, you can't do either of these things unless the iron is "there," available to be used. The human use of a material involves some interest in its origin and history, that is to say, in a series of events which may have occurred long before any knowledge supervened upon them. Pragmatists have a Philistine taste for immediate satisfactions, which makes them take more interest in cash in hand (the metaphor is Jamesian) than in the complex system by which cash becomes cash.

Positivism has a more scholarly (or at any rate a less economic) reason, which is that you can arrange no test to determine whether anything exists without being known. If I want to find out whether there are thirty students in my class, I can count them: that's a test. If I want to know whether five of them made A's, I can look in my rollbook: that also is a test. But if I want to know whether the thirty students exist independently of my knowing them, I am, say the positivists, helpless: there is no test for that. Any test would immediately connect them with my awareness of them or somebody else's awareness of them, and the crucial negative case (John exists, but at that moment is not being known) would always escape examination.

This argument seems a good deal more powerful than it is. There are surely very few teachers who seriously doubt the independent existence of their students, nor would they believe the question "meaningless" because untestable. Moreover, quite a number of such questions can be found. For example, suppose you wanted to prove empirically that the stars do not determine the course of human events. The obvious way would be to obliterate the stars, and then show that human events occur as before. But the stars can't be obliterated, and hence the negative case must remain unobtainable. Yet this doesn't seem to prevent us from regarding astrology as positively superstitious.

The reason lies in the weight of empirical evidence

which we do have. Psychological, sociological, and historical causes go so far to explain the course of human events that there is really nothing left over for astrology. And so it is with our question here. The hypothesis that things exist independently of our perceiving them goes so far to explain their behavior that we really don't need any other hypothesis.

The significance of the question, however, best appears when we see it in terms of society and social struggles. If you arrange the various philosophies from right to left according to their political implications, you will find that positivism and pragmatism belong to the liberal center; historically they descend from the English empiricist tradition, which stretches far back into the Middle Ages. The right wing is constituted of transcendental philosophies, and the left wing of materialist philosophies. Now, both these groups take very seriously the question whether things exist independently of "mind." The right wing insists that you must believe in a supernatural; the left wing, that you must not. Both, however, agree that the political implications are profound. Only the liberals think it doesn't matter.

We could multiply examples, but here is a recent one, stating the right-wing view and contained in an AP dispatch from Vatican City:

> People who do not believe in God can only be held together by terror, Pope Pius XII told Rome university students yesterday [June 15, 1952].
> Speaking to 2400 students and professors, the pontiff said fraternal order in modern society can only be on the basis of Christian principles.
> "A palpitating proof of this," the pontiff said, "is (offered

by) masses which do not believe in God and which can only be held together with terror."

The Pope saw in this "a clear proof of the existence of God, and of the beneficial repercussions that the belief in Him gives to men."[1]

And as an example of the left-wing view, we may take the celebrated passage in which Marx contrasts himself with Hegel:

> To Hegel, the life-process of the human brain, *i.e.* the process of thinking, which, under the name of "the Idea," he even transforms into an independent subject, is the demiurgos of the real world, and the real world is only the external, phenomenal form of "the Idea." With me, on the contrary, the ideal is nothing else than the material world reflected by the human mind, and translated into forms of thought.
>
> ... The mystification which dialectic suffers in Hegel's hands, by no means prevents him from being the first to present its general form of working in a comprehensive and conscious manner. With him it is standing on its head. It must be turned right side up again, if you would discover the rational kernel within the mystical shell.
>
> In its mystical form, dialectic became the fashion in Germany, because it seemed to transfigure and to glorify the existing state of things. In its rational form it is a scandal and abomination to bourgeoisdom and its doctrinaire professors ... because it regards every historically developed social form as in fluid movement, and therefore takes into account its transient nature not less than its momentary existence; because it lets nothing impose upon it, and is in its essence critical and revolutionary.[2]

---

[1] The reader must discount the absurd tone which a barbarous translation has imposed upon the Pope's language. In any case, the point here is not the truth of the remarks but the fact that they draw political implications from a belief in the supernatural.

[2] From the ending of Marx's Preface to *Capital*.

If these two quotations are read in the present historical context, it will be plain that the Pope and Marx, though they disagree about the existence of a supernatural order, agree perfectly with regard to the political implications: viz. that if you are a metaphysical idealist, you will tend to be a political conservative; and that if you are a metaphysical materialist, you will tend to be a political radical.[3]

Now, it seems to me that if two parties to a profound social conflict regard a certain philosophical problem as significant, then it *is* significant. Indifference to the problem suggests a wish to stay aloof from the conflict or even to deny its existence. But that, in its turn, is a grudging, inverted confession that the problem is significant, since aloofness is terrified interest. Pragmatists and positivists see the problem much too narrowly: the one, as irrelevant to immediate concerns; the other, as escaping experiment. But, seen in the whole context, as philosophical ideas ought always to be seen, the problem is clearly so significant that it is bound up with the entire fate of future society.

We have a real problem, then. Let us look at it.

[3] I use the term "metaphysical" to indicate that "idealist" and "materialist" have in this context no ethical connotations whatever. An unlucky linguistic accident has allowed "idealist" to mean two quite different things: (1) a man who believes that the universe is fundamentally spiritual, and (2) a man dedicated to the service of moral ideals. Similarly, "materialist" may mean (1) a man who believes that the universe is made up not of spirit but of matter and energy, or (2) a self-seeking, anti-social scoundrel. Obviously the two sets of meanings are quite different: a metaphysical idealist can be (and not infrequently is) a moral materialist; and a metaphysical materialist, needing to live down the opprobrium of his name, is almost always a moral idealist.

Esse Est Percipi

The idealist position is that "existing" and "being known" are equivalent terms. I imagine that of the hundreds of thinkers who have held this doctrine few have meant that an object, in order to exist, must be *completely* known. Hegel seems to have meant this, but not many others are so rigorous in their demands. What is generally required is that the object be known in at least one of its aspects. The refinements which Berkeley introduced upon the doctrine, and which remain standard, restrict the sense of "being known" (and, consequently, of "exist") to simple presence in consciousness. In this form the doctrine is least vulnerable, for it has the least number of assumptions. And since this is the form in which we shall examine the theory, perhaps we may allow ourselves the use of Berkeley's Latin tag: *esse est percipi,* "to be is to be perceived."

Now, if *esse* is really *percipi,* somebody or something must do the perceiving. That is to say, there must be human consciousnesses, or divine consciousnesses, or one cosmic consciousness blandly sustaining the universe. Either all this, or else nothing exists, or *esse* is not *percipi.*

We shall set aside for the moment the last two alternatives, and so we are left with the first three. Now, what about the existence of these consciousnesses? Does that in turn depend upon being perceived? If so, the doctrine falls into an infinite regress, like boxes containing other boxes with never a box to contain them all. For if my table exists only because I perceive it, and I exist only because some other consciousness perceives me, and so on toward infinity, we shall all collapse unless we can rest on some being whose *esse* isn't *percipi*, and who (or which) can exist without the aid of any other consciousness than its own.

This infinite regress has been universally rejected, although it is not, perhaps, logically impossible. Consequently, the philosophers who support idealism will admit that there is at least one being for which "existing" and "being known" are not equivalent terms. The history of philosophy contains a lot of these beings. Berkeley had himself, other people, and God. (He might have added animals, but didn't.) Hume had sequences of sensations. Hegel had the Absolute Mind. James had streams of consciousness. The positivists, profoundly sunk in individuality, have got each himself. Obviously, among the very professors of the doctrine there is a radical belief that in some sense, at any rate, existing and being known are two different things.

This fact may seem a crusher, but its only effect is to limit the theory. No one seriously supposes that his own existence depends upon somebody else's being aware of him. But what about the inanimate world of stones and trees and oceans and glaciers, which cannot answer back as we can and proclaim their own independent existence? It may be, for all we have thus far shown, that this world does exist solely because, and in so far as, we perceive it. This would be some sort of triumph of "mind" over "matter," and it was precisely upon such grounds that Berkeley announced himself the final vanquisher of materialism.

What can be done with this new assertion? We now say that beings capable of perceiving and knowing do exist independently and in their own right, but that the world of objects, of bodies — in short, the physical world — exists only so far as the perceivers and knowers have it in their consciousnesses.

Certain perplexities at once appear. For example, the earth is vastly older than man, and the universe is older than either. It cannot have been *human* consciousness which kept the universe in existence before the arrival of man. Consequently, we must either reject the current anthropology and geology or we must postulate the existence of a nonhuman consciousness.

Again: the light I now see from a certain star left there two thousand years ago. There was no human observer when it left, and there has been no human observer until now. Consequently, I must either reject what astronomy has to say in this case or I must postulate the existence of a nonhuman consciousness.

Again: the center of the earth is not, and never has been, the immediate object of anybody's awareness. Consequently, as before, we must either suppose that no such center exists or that some nonhuman consciousness maintains it in existence.

We are now in a position to state the principle which these examples illustrate. A great deal of science has to do with events occurring quite outside the reach of human observation. If, nevertheless, to be is to be perceived, we must postulate some sort of consciousness in which these events can occur. That consciousness will not be human consciousness, nor, indeed, any kind of animal consciousness. It has to be a consciousness (or consciousnesses) capable of vast reaches through space and vast endurance through time. It has to be the consciousness of a God, or of gods, or of a world soul. The transcendental consciousness thus deduced is capable of experience and to that extent resembles ours. Otherwise it is radically different. There is no direct evidence for its existence which we can get from introspection into our own lives or from observation of other people's behavior. The case seems rather to be that idealism assumes the existence of such a consciousness in order to avoid the absurdity of rejecting science. Granting, that is to say, the principle of *esse est percipi*, a defense of science drives you to the supernatural.

Now, there is nothing in the argument to show that science cannot stand on its own feet. The drive toward the supernatural evidently comes, not from science, but from idealism. The reason is that consciousness, as idealists conceive it, is a supernatural entity even on the human level. For by the phrase "human consciousness" idealists don't mean a brain and nervous system acted on by other material objects through the sense organs. They mean a nonmaterial something, a mental or spiritual entity, which receives sensations and conducts an activity called "thinking."

This notion is the vestigial remnant of a grander idea. In the Middle Ages man was conceived to be an immortal soul pilgrimaging between eternities. With this concept, sublimely elaborated, the feudal lords hid the historical fact that for most men they made the pilgrimage far from pleasant. The concept, bequeathed even to our own time, still lives among the laity; but on somewhat more sophisticated levels it has undergone interesting transformations. In the seventeenth century, for example, "soul" became "mind" — a shrewd, cognizing entity which had largely lost its heavenly destination and busied itself with profitable inquiry into the physical world. A man of commerce doesn't need a soul, for his treasures are laid up where moth and rust doth corrupt and where thieves break through and steal. But he must analyze and calculate, and so needs a mind.

This concept too survives, side by side with the medieval, which indeed it now matches in honor. Note the tone of the concept as Mr. Dean Acheson used it in a speech before the American Society of Newspaper Editors, April 22, 1950:

> The adventurous people who settled the eastern shores of North America in the seventeenth and eighteenth centuries brought with them certain ideas which had come down to them through the whole stormy history of civilization. The first of those ideas was freedom — freedom of the mind and spirit, the most adventurous and dynamic idea ever to seize the mind of man. It drove men — and it continues to drive men — to inquire into the relation between man and God; to study the nature of the universe; to explore the purposes of human society.
>
> Every thought we have in our minds, every relationship we have in our private lives, every institution under which we live, all of modern science has been molded, and in many cases created, by the exercise of the freedom of the human mind.[4]

At least half the actual causation is missing from this account, for there is no mention of the effect of physical nature and social institutions upon mind and the freedom thereof. But that only shows how honorific the

---

[4] "The Security and Well-Being of Our Country," in *Vital Speeches*, Vol. 16, p. 453.

concept of mind has become. With mind (and perhaps some cash) you can, it is supposed, do anything.

During the eighteenth century, "mind," a shrunken form of "soul," began to shrink in its turn. It became "consciousness," that is to say, mere and simple awareness of sensations. No doubt this was due, among philosophers, to a wish to make as few assumptions as possible. But I find it hard to resist the idea that mercantile capitalists, having got a monopoly of "mind," preferred the rest of humanity to have nothing more than awareness. From an advertiser's point of view consumers need to have awareness, but they become too critical if they have anything more. They need to be aware of the letters LS/MFT, and it is permissible for them to be aware that LS/MFT stands for "Lucky Strike Means Fine Tobacco." But the presence of any more analytic faculty would break the charm, with who knows what disasters for the sales office.

"Consciousness," though it sounds agreeably scientific, has never had the flattering tone of "mind" or "soul." It suggests too little. You cannot compliment a man by telling him he is conscious, though he will beam with pleasure if you tell him he has a mind. James was quite right in describing consciousness as "the faint rumor left behind by the disappearing 'soul' upon the air of philosophy." [5] Yet it remains a favorite concept with philosophers, and, as we see, is quietly used in a practical way.

The soul, as capable of surviving bodily death, was clearly something supernatural. Its successors inherited the same attribute. Mind and consciousness are, both

[5] *Essays in Radical Empiricism*, London, Longmans Green, 1912, p. 2.

of them, something more than brain and nervous system. They escape measurement in space and time, and they aren't observable by any sense organ. You are, it is true, supposed to "see" them when you "look into yourself," but it is very doubtful whether this sort of evidence shows anything more than movements of physiological tissue.

The conclusion is that all metaphysical idealism — of which the Hegelian type is a spectacular example — rests upon a belief in the supernatural. We have now to discuss the validity of this belief.

A Farewell to Supernature

The concept of things supernatural no longer stands in very high repute. All of us have, from time to time, got rid of various portions of it — of Santa Claus, of ghosts, or of witches. In modern thought it has been subjected to tremendous criticism, and no doubt that is why philosophers ordinarily allow themselves only that little bit of it which the term "consciousness" conceals.

There are three sorts of evidence against the existence of a supernatural order. They don't, to be sure, demonstrate it to be impossible, but they do render it extremely doubtful. When you come to make up your mind on the question, you have to decide between these doubts and the bare possibility that the doubts are mistaken.

(1) If the supernatural exists, it must be in some degree knowable or else we cannot know it to exist. If it is knowable, the knowledge of it will be relevant to other knowledge we have. That is to say, knowledge of the supernatural will be needed to complete our account

of the world and will modify our knowledge of the natural.

This being the case, it is very singular that science, which is the norm of human knowledge, makes no use of the supernatural whatever. The average textbook on physics has nothing to say about God. The average textbook on psychology has nothing to say about the soul. Throughout the whole of science the host of traditionally supernatural beings — gods, angels, demons, and spirits of all degrees — are nowhere to be found.

The sciences, indeed, appear to have grown just in proportion as they left the supernatural amid the mists and magic of their origins. Take any of the great scientific generalizations like relativity, evolution, or gravitation, and add to it some appropriate doctrine about the supernatural. Then ask yourself whether the added doctrine explains anything which was not already explained by the generalization. You will find that it does not.[6] In science the supernatural is always excess baggage, which you may carry for comfort or importance, but which you do not otherwise need.

(2) It is extremely difficult to make out just what the supernatural, if it exists, contains. One can get at the natural world by sense experience and scientific method, but to the supernatural neither of these gives any access at all. Logic, divorced from scientific method, seems not to help very much: it can only make sure that the supposed contents of the supernatural do not contradict one another.

Perhaps this is why mystics, the most faithful geog-

[6] For example, Newton didn't need to consider space as *sensorium Dei*, "God's sense-field," in order to explain the phenomena he was

raphers of that region, reject logic entirely. They prefer to judge by the inward eye, which sees what it sees and claims many a fair vision. But this sort of belief is bought at the price of rational ideas. Suspicion grows, therefore, that the supernatural is a world of fantasy and feeling, of "dreams that wave before the half-shut eye." It is a world which might be, but cannot quite persuade us that it is.

(3) There is much reason to think that belief in the supernatural rests, not on proof, but on social conditions. Suppose, for example, that scientists were to say, "We think that the universe as science describes it is all the universe there is." And suppose that, working out the implications of this concept, they went on to say, "We think, therefore, that there are no gods or demons, nor any immortal souls, nor any resurrection of bodies, nor any recompense for present sufferings in an afterlife."

Suppose all this, and you will soon see that much more would happen besides scientists being hounded out of their jobs. A great many people would feel that, if the universe is really of such a sort, then life is bare, hard, and futile. To millions of men God is all that an earthly boss is not: kind, fair, and free of piousness. To millions of men human immortality offers the hope of success in a divine economy more rational and more just than any they have ever known.

Simply as an eclipse of certain human hopes, the sort of universe described in the sciences makes an enormous difference. In the gloom of that eclipse (which would be

---

analyzing. My whole argument in this paragraph is borrowed from Spinoza's attack on "final causes," *Ethics*, Part I, Appendix.

temporary, as all such darkenings are) vast stirrings would shake the fabric of society. For men, persuaded that they have but the here and now, the one chance, the single journey within time and space, would strive to make that journey joyous, that chance successful, that here and now bright with all possible bliss. Having no supernatural power to lean upon, they would turn for strength to one another and end the schisms which, dividing them, divide also their talents. And if they came too early in time to enjoy these fruits themselves, they would seek them, willingly enough, for their children.

Let us put these predictions within one example. Suppose it were possible, as it now very nearly is, to prolong physical life indefinitely by medical care, good diet, and freedom from shock. Would a man having such means to such an end at his disposal set them all aside in favor of the immortality promised him by various religions? Of course not. The generality of men already try to live as long as they can, and use every available means to do so.

In this as in other respects men use the here and now to maintain the here and now. Their applications to the supernatural are clearly a last resort. They are, one and all, like the old lady, during a great storm at sea, who, being told by the captain that there was nothing to do but pray, exclaimed, "My God, is it as bad as that?"

There is little doubt what men would do if they were entirely convinced that the universe of space and time is the only universe. But in their struggles to make that universe blossom with fruits for present plucking, they would collide with some few other men whose purpose is to keep all plucking to themselves. These would be (and are) the exploiters and profit makers, the cartelists and colonial tycoons, whose way of life has long been a system of theft modified by murder.

Suppose, before the moment of collision, the many say to the few, "Our notion is that we have this one life in this one world. Share with us therefore the labor and the fruit, and let us all be happy." How can the few, in good conscience, refuse so equitable a demand? Now suppose the many, instead of taking their stand on science, say, "Our notion is that men have two lives in as many worlds. We admit that in one of these lives and worlds men are to toil without much happiness, while in the other they are to be happy without toil. Nevertheless we think you might share the bounties of earth with us much more freely than you do."

Could not the few reply, and have they not been replying for some centuries, "You have no real complaint in the end. The other world, which requites your sufferings, demands of you meanwhile, as the sign of your desert, that you work faithfully for us"?

Obviously, in respect both of ethics and of propaganda, the profit makers have stronger ground in this second philosophy than in the first. They need the supernatural as they need the air they breathe: it is one of their most powerful weapons for the prevention of revolt. Their remarkable concern for religion since 1945 is not, as it may have seemed, a deathbed repentance. It is an effort to make the deathbed impossible and the repentance unnecessary.[7]

---

[7] One of the brassiest of these appeals to religion will be found in a speech by Mr. Ralph Chaplin before the Controllers Institute of

In view of the three sorts of evidence we have just examined, we are, I think, entitled to conclude that the supernatural is not a scientific concept but a philosophic illusion having certain social origins and purposes. It cannot be finally destroyed by criticism alone, and it will survive so long as the social need for it survives. But when that need has vanished, the supernatural will wither away like the State.

The Paradoxes of Change

The argument thus far developed limits real existence to the world of space and time. In this world everything is a happening, an event; that is to say, it is either a process, or a moment in a process, or both. For example, a twenty-five-year-old oak tree *is* the whole process which constitutes its growth, and also it *is* that particular oak tree at the moment it reaches twenty-five years. This is why everything in the universe can be described genetically (according to its derivation) or analytically (according to its structure). And so, of course, can the universe itself.

Now, change, familiar enough as an observed fact, appears extremely paradoxical when you come to analyze it; and this paradoxical aspect has been, from the time

---

America, September 26, 1949: "As a preliminary step in insuring industrial peace I would suggest in all sincerity that the Chairman of the Board and the top officer of the union hold themselves personally responsible for industrial peace, and that two chairs be left vacant between them — one for John Q. Public, and the other for the divine Author of the Sermon on the Mount. Industrial relations are the Achilles' heel of a free economy. If we fail to find the answer, the freedom of man and the Church of God will join Babylon and Tyre in the ashcan of human history." (*Vital Speeches*, Vol. 16, p. 218.)

of Parmenides, the philosophical ground for rejecting it altogether. No doubt a rejection of sense experience follows too, but for believers in a static and purely structural logic such a rejection has no terrors. Conversely, as we saw with James, the devotees of immediate experience are often willing to throw logic away if it seems an obstruction to activity.

Happily, the supposed irreconcilability of logic and sense experience arises from a misconception of each. The older, Aristotelian logic is valid as far as it goes, just as Euclidean geometry is valid as far as it goes. But a mathematical account of change requires calculus, and similarly a logical account of change requires dialectics. The rest of the chapter is an extended commentary upon this assertion.

We shall begin by stating some of the paradoxes in the concept of change.

(1) Change seems to require that an object be itself and at the same time not itself (since it is becoming something else). When the oak tree was a sapling, it was a sapling; but it was also not a sapling, in the sense that it was on the way to full growth. The ice cube in your tumbler at dinner is an ice cube; it is also not an ice cube, in the sense that it is melting into water.

This condition appears to violate the law of identity and thus to introduce an irrational element into the world. However, if you are willing to entertain a brief technical argument, you will soon see that the law is not violated after all.

The law of identity asserts that everything in the universe is just what it is, so that of any entity A we can always say, "A is A." Now, if A signifies a process, then the law certainly holds. Every process is exactly what it is: the process, for example, by which a particular acorn became a sapling and that sapling became an oak is just the process it was, and no other. "A is A."

Or, again, if you wish to consider any stage of a process by itself, the same thing holds. The oak tree aged twenty-five years is that particular oak tree aged twenty-five years, and no other. "A is A."

Irrationality is not in change at all, but is in the minds of some observers. What you find in these "correctors" of Aristotle is a whimsical manipulation of terms. They tell you, very sagely, that the oak tree as sapling is not the oak tree aged twenty-five years. Now, if "A" stands for the oak tree as sapling, then "A" cannot in the same context stand for the oak tree aged twenty-five years, any more than "eagle" can in one and the same context stand for a live bird and a standard carried by Roman soldiers. When we are told that a sapling oak is not a twenty-five-year-old oak, we are really being told that A is not B. But I think we knew this in the first place.

Perplexity in the concept of change doesn't lie here. But perhaps it does lie in another paradox, which is our second.

(2) It is the nature of change to have states, of a process to have stages. One of Shakespeare's most famous passages describes the "seven ages of man," that is to say, seven characteristic states in the growth and decay of human beings. Within a process these states (or stages) are spread out successively in time, one following another.

Now, the relation between any two stages does seem to be of a curious sort. (*a*) The stages must be continuous, for otherwise all connection between them will fail and the process itself will collapse into isolated fragments. But (*b*) the stages must be different from each other, or else they won't be distinguishable: if the oak tree as sapling isn't different from the oak tree at twenty-five years, you won't be able to tell them apart. To the extent, however, that the stages are different, there is *discontinuity* in the process.

Perhaps you have heard of the baronet (why baronet I don't know, but in this story he always is) who had a pair of silk socks which, over a space of years, were darned away entirely into wool. If I ask you whether the socks at the end of the darning process were the same as at the start of it, your first answer will be negative, because you are impressed with the change from silk to wool. Nevertheless, the baronet, if we asked *him*, could go to his bureau drawer and produce the socks. And this he could not do unless the socks were, in some sense, the same.

The puzzle, if one exists here, can be resolved as follows: The "sameness" of the socks lies in the continuity of the darning process through which they have passed. The difference in the socks lies in the discontinuity among the stages, some of which had less darning and some more. Unless we regard the process as both continuous and discontinuous, we seem to get no adequate or even intelligible description of it.

The kind of logic called "dialectical" accepts this account as valid for change in general. Continuity and discontinuity it holds to be, not flat contradictories (the one eliminating the other), but opposites whose interplay makes the movement of events the sort of movement it is. Furthermore, it holds that this interplay accounts for a very striking fact, mentioned with singular rarity by philosophers, namely, that time is irreversible, that it never goes backward but only forward. Presumably this is because every latest stage contains something genuinely new, which, though the product of past causes, thrusts those causes definitively behind it. There is always something new under the sun. For that matter, the sun, as Heraclitus said, "is new every day."

(3) The problem of relations among the stages of a process is part of the general problem of relations, and the special dialectic of the one derives from the larger dialectic of the other. This gives us our third paradox.

Take any relation at all having certain members in it. Does the relation affect the members? Do the members affect the relation? Do the members affect one another?

Let the relation be, for example, marriage. Setting aside all cynical conjecture as to which spousal member is ordinarily dominant, we shall have to say that the answer to all three foregoing questions is, Yes. For, unless my observation of the married state is hopelessly erroneous, the relation does affect the spouses (it *makes* them spouses, in fact), the spouses do affect the relation, and the spouses affect each other.

Now let the relation be one of position in space: A is to the west of B. This is a much looser relation. If we limit discussion to space alone, A may not seem much altered by being to the west of B. Nevertheless it *is* somewhat altered, for it now has a quality it wouldn't otherwise have, namely, of being to the west of B. And if the spatial relation is involved with other relations, the differences grow more marked: for example, if *A* and *B* are cities fifty miles apart, your journey from *A* to *B* will differ tremendously as you take an eastward or a westward route. It seems that for the looser relations of space the three questions still get the same answer, Yes.

At this point, however, the puzzle begins. If the relation *completely* determines the members, then the members, lacking any individuality, are ciphers, and the relation relates nothing. On the other hand, if the members are granite, immutable entities, they cannot affect one another, and the relation becomes nothing. (Thus, as we previously observed, if the stages of a process are completely discontinuous, there won't be any process; and if the process is completely continuous, there won't be any stages.)

Evidently there are two extremes, either of which renders relations inexplicable: relations may be conceived as obliterating their members ("the night," as Hegel once said, "in which all cows are black"), or the members may be conceived as shattering the relation. These extremes have in fact been militantly advocated, and the second of them is even now very probably the dominant view. They used to be known, respectively, as the theory of internal relations and the theory of external relations, where "internal" means that the relation makes the members what they are, and "external" means that the relation is only incidental to the members. It was supposed that you had to choose one view or the other.

Dialectics, however, follows Plato's advice to behave like children and say, "Give us both!" As before, we shall take the individuality of the members and the integrity of the relation, not as contradictories, but as opposites modifying each other. And thus we seem once more to get a logic capable of stating the structure of change.

Before leaving the subject, it may be worth while to consider how this theory resolves the old paradox of free will versus determinism. This problem turns on how you conceive the relation between a human being and his environment. The extreme advocates of free will maintain that in every human being there is something (usually a faculty of choice) not influenceable by environment. The extreme determinists maintain that environment determines human behavior completely.

The free-will view is, quite clearly, a form of the theory of external relations; and the determinist view is a form of the theory of internal relations. We have, however, dismissed these two theories as extreme distortions of the actual state of affairs, and we have said that *all* relations are both internal and external. Consequently the relations between man and his environment are both internal and external, also; that is to say, man and environment modify each other. To the extent that men, singly and collectively, modify their environment, they are "free"; to the extent that their environment modifies them, they are determined. It seems plain enough that men aren't absolutely free or absolutely determined. A study of their evolving history will show how much they are the one or the other.

Strange to say, men need the determinateness of the world (and of their own nature) in order to be free. Their ability to satisfy their wishes is in proportion to their technology, their technology is in proportion to their knowledge, and their knowledge depends upon regularity of pattern in events. At the same time, events do get altered by human action, and therefore reflect the "free" causality of man. Thus free will and determinism are theories which are not only compatible but necessary to each other. Taken dialectically, they together perfect one's vision of the world.

Thales to Heraclitus

In ridding change of its paradoxes, we have stumbled upon not a few suggestions as to what change is. These suggestions need now to be generalized into a full description. To do this is to recapitulate the history of early Greek philosophy, and I have the feeling that it may be interesting as well as profitable to carry out the task in those terms. What we are particularly to watch is the triumph of Heraclitus over Thales, of dialectics over the concept of substance. For this triumph there is a modern parallel: the hypothesis that matter and energy are ultimately identical has replaced the old "hard atom" of nineteenth-century science.

Thales of Miletus, the first Western philosopher, thought that water was the primordial stuff of the universe. This assertion, not so trivial as it may seem, is in any case less important than the question it sought to solve. Thinking on analogy with human technique, Thales had apparently asked himself, "What is it all made of?" The question assumes that the universe gets developed, like pottery, from a common material. Thales wanted to know what that material was. He didn't ask — or at least we don't know that he asked — why things change or how there comes to be such variety in the world. He was looking for origins, and he took those origins to lie in one universal self-identical stuff.

Thus was born the concept of substance, the oldest and until recently the most respectable in philosophy. In all that time a great deal has been written about it, but not much has really been said about it. This is because you find, when you come to the heart of the subject, that hardly anything can be said. What you say about any entity amounts, in the end, to listing its qualities, its quantities, and the relations in which it stands. But all these are different from substance itself. You are therefore in the position of trying to describe substance in terms which are not substantial, and then it appears that you aren't talking about substance at all.

This indescribable entity called "substance" can be conceived as cosmic stuff or as elementary particles, but in either case it is inwardly changeless. It can take on qualities and relations as a body takes on clothes, without for one moment ceasing to be itself.

How primitive and "natural" this conception is may be seen in the syntax of languages (for syntax is naïvely philosophical). We say, "That coat is brown," "My gloves are on the right-hand side of the table." As in these examples, qualities and relations have a habit of appearing in the predicates of sentences. What is it, then, that serves as subject? Well, perhaps, substance. Aristotle actually defined substance in this way, as that which is always a subject and never a predicate. Though we are forever ascribing to it various qualities, quantities, and relations, we rarely get a chance to say what the thing itself is; and then, it appears, we are safest with a grammatical definition. Within fairly narrow limits this scheme is workable; at least, the human race has not been extinguished by such confinement. But the impenetrability of substance — its intransigeance, so to say — plays hob with every theory based upon it. The prime lesson of sense experience is that things change. Consequently, a separation begins — at first a rent and then a gulf — between the changeless substance and the changing world.

To the extent that substance dominates philosophy, no solution to the paradox of permanence and change can be found. Nor will it suffice, in the Jamesian manner, to replace the dull lingering of substance with a disorder of flux. Permanence and change are two concepts which we must try to find, like lovers, in a common embrace, exposing thus no unnatural secret but rather searching out the unity amid all variance and the concord amid all strife.

While the Ionian sun was rising toward its zenith and Miletus still led the cities in philosophy, there appeared the rudiments of a theory which was to solve the paradox. Anaximander detected in the universe a reconciliation of change and order. The reconciliation seemed, indeed, so apt that his language, in expressing it, took on a moral tone. Events, he said, suffer punishment and make reparation to one another for the "injustices" they commit in the flow of time. Each process tends, as it were, to excess, and thus evokes a contrary process for its own modification.

This description of change as the interplay of opposites is not limited to ethics, but applies to the universe generally. We can think of the interplay as establishing a pattern and of the pattern as surviving upon the very back of change. Consequently we can say that it is the diversity of things which gives them unity, it is their movement which gives them form. On this view, the permanent is no secret substrate behind the floating face of things, no refuge from the hounds of circumstance; but it is the form and consequence of change, in which it rests and prospers.

The dialectical theory, thus born, reached its first manhood about 500 B.C., in Heraclitus of Ephesus. The genius of this man speaks to us through some six score fragments, the number of which, though far less than we should wish to have, shows him to have been the most quotable of early philosophers. He performed, indeed, the most difficult of philosophical feats: a persuasive revelation of the universe as it is, stated with a voice indubitably his own.

"Much learning does not necessarily teach understanding." [8] With this ironic comment Heraclitus dismissed the polymaths, his predecessors. Now, understanding requires completeness of description, and completeness of description requires that any object be observed in its context, its environment — that is to say, in terms of its "opposite," of what it is not. Accordingly, Heraclitus says, "God [i.e., the totality of things] is day-night, winter-summer, war-peace, satiety-hunger; but he takes on different forms, just as fire, when mingled

[8] Fragment 40 in Diels's numbering: *Die Fragmente der Vorsokratiker*, Berlin, Weidmannsche Buchhandlung, 1912, p. 86.

with various kinds of incense, is named according to the odor of each." [9]

If movement is an essential condition of things, the opposition between any one entity and its environment is a dynamic affair. Things modify one another; there is a certain "strife" among them. From these interactions nothing withdraws unaltered, but everything is different from what it was and is different because of the interactions. This is what makes the movement move and the process proceed. Heraclitus rebukes Homer for wishing that strife might vanish from the universe: if strife, in the general sense of interaction, were to vanish, the universe would vanish with it.

This doctrine of the strife of opposites has remained a basic part of dialectical theory. It involves, as Heraclitus also saw, a second doctrine: the unity of opposites. For "opposition unites, and out of discord comes the loveliest harmony"; there is an "attunement of opposite tensions, like that of the bow and the lyre." [10] The literal sentence behind these metaphors is that entities which interact are united while they do so, and the unity is what it is because of the effects the entities are producing in each other.

We can illustrate the principle quite simply. Heraclitus reminds us that a potion will separate into its ingredients if it is not stirred. When next you bring

---

[9] Fragment 67. The English translations usually have "and" between the nouns, e.g., ". . . is day and night . . ." The conjunction, however, is not in the original, and I surmise that Heraclitus intended to suggest a peculiar intimacy in the relation of these opposites, such as might better be indicated by a hyphen.
[10] Fragments 8 and 51 respectively.

from the pharmacy a bottle of liquid medicine bearing upon its label the legend SHAKE WELL BEFORE USING, you can pause to reflect how good a dialectician your doctor was. For the essential and effective nature of the medicine (its "unity") is established and maintained by a certain movement among its parts. Without that movement, the parts separate, the unity is lost, and therapy fails.

Thus the strife of opposites is the source of movement, and the unity of opposites is the source of pattern and form. The strife and the unity, taken together, yield a third principle of dialectics: that change is passage into novelty. The alterations which occur when things act upon one another are events that never existed before. No doubt these events still show the evident marks of their predecessors, but they also show some part, some aspect, some structure which is new.

The older science regarded change as repetitive, like the apparent behavior of machines. Yet, when the engine of your motorcar dies upon the highway, you realize that its thousand cycles of movement, which seemed so safely identical one with another, had in fact more differences than you thought. Each cycle produced some infinitesimal novelty, and the accumulated novelties erupted at last into spectacular change. The mechanistic description is partly true, for there are resemblances and the resemblances are often very close. But a man who thinks change to be nothing more than mechanical may find his life collapsing under the shock of what is new.

Thales to Heraclitus — Modern Version

Philosophy was in the arms of Thales, when, in the fall of 1923, I began the study of it. Looking back upon that introductory course, wherein the problems of philosophy were displayed in staggering size and number, I can see very well that substance was lord of the universe. We sat, five hundred sophomores in a required course, while Professor Spaulding went through the list of problems — would they never, we asked ourselves, end? The Ontological Problem, the Cosmological Problem, the Teleological Problem, the Epistemological Problem, the Axiological Problem, and so on to the exhaustion of pencil and mind.

And there, right in the beginning, in the Ontological Problem, was substance, with due meditation upon mind-stuff and body-stuff. Were you a monist? Then you thought there was just one substance. Were you a dualist? Then you thought there were two substances. Were you a pluralist? Then you thought there were many. And you went to lunch, to gym, to supper, to the nightly battle of the books, wondering whether food, sneakers, and pages were all of one substance and that substance identical with the mind which absorbed, digested, or repelled.

And there, in the Axiological Problem, was substance also. For axiology (so you learned) was the theory of value, and the theory told you (among many other things) that values have to have some mode of existence during those unfortunately frequent intervals when flesh and blood don't possess them. That mode had to be existence "in" something, and for this purpose you needed a kind of substance — spiritual, no doubt — which would offer secure residence for values.

Thus substance appeared a scientific and an ethical necessity. Around it all your loves and yearnings wove themselves, in the familiar manner of men. For who, at eighteen, will confess that he admires body and behavior more than "soul"? Behaviorists might prosper, as they later did, only to be thrust aside. We, for our part, were monists with one spiritual substance, or dualists with a preference for mind, or pluralists with a little bit of soul in everything. It is the way you feel when you are eating regularly.

It was the way you felt in the prosperous Coolidge times. Strange, bitter-sweet times they were for intellectuals, whose values were constantly abused by a horde of profit seekers, chanting, as the market rose, the insuperable excellence of practical success. Some of us fled to civilization as the Left Bank knew it; all of us read, month by month, with mounting snobbery, the Americana column in the *Mercury*. The world, it seemed quite obvious, was too much with us, and we laid waste our little powers — not, however, with getting and spending, but with grief over genius ignored. And we fiercely resolved that our admired values — truth, goodness, beauty (for so we named them) — though scorned in office and in market place, must nevertheless survive, impalpable and unseen, behind some bastion of the spirit.

Thus moral idealism, thwarted by the vulgarities of commerce, made us metaphysical idealists. It was an easy solution and apparently virtuous. Perhaps we should have known that easy virtue is no virtue at all, that our best service for beauty and truth would have been to make the world fit for them to inhabit. Perhaps we should have known that philosophers turn their frustrations into idealism as poets turn theirs into verse. But behind the mist of metaphor (taken, alas, as sober fact) there grew a dark awareness that the values of humanity and the values of commerce were somehow in conflict, and that beauty faded wherever profits bloomed. Then, after 1929, the veil of metaphor was lifted, and we perceived that, although every man wants to eat, there are some men who are willing that many should starve.

Before the social realities broke in upon us, philosophy of the 1920's had much the liveliness of a Breughel landscape. Of our conceptual trinity, goodness had fled to dwell (not very comfortably) on the Hegelian monolith, beauty was aloft presiding over Santayana's realm of essence, and truth was wandering in the forests of methodology. Pragmatists were to be observed (in the foreground, of course) shouting, running, abetting every activity; while realists were studying objects through those lucid windowpanes which symbolized, in their theory, a consciousness washed clean of personal bias.

The chief philosophical concern in those days, as again now, was with the theory of knowledge. With this problem we were much at ease. We engaged interminable disputes about perception; and if we ever stole out of the Ego into the universe, it was to inquire whether the universe was friendly.

With the death of Bosanquet in 1923 and of Bradley in 1924, metaphysics itself seemed to have died. The pragmatists had been digging its grave for three decades and had announced its death frequently though prematurely. Now all was over. We would let the scientists tell us what universe they knew, and we would tell the scientists how they knew it. It was a nice division of labor. The trouble was that neither group could do its own work without doing some of the other's.

Then in 1929 came Whitehead's *Process and Reality,* couched in a vocabulary entirely new and formidably difficult, so that for the first time since Hegel philosophers had to learn words as well as ideas. It was obvious, moreover, that metaphysics had risen from the dead, not as a whisper in the midst of methodology, but in its old majestic form of free speculation about the cosmos. All the little Egos cracked open like eggs; those that were fertile prospered, and the others dried quickly in that wholesome air.

The man who wrought all this was enormously talented for the purpose. A mathematician by training, he had mastered also the chief concepts of physics and biology. The struggle of these concepts within his mind must have been lively, but in the end it was the biological ones which triumphed. They did so because he thought that change could be better described on analogy with *growth* rather than with mechanical pushings. He considered, moreover, that this notion would successfully unite relativity theory with the phenomena of sense experience, and thus avoid what he called, in tones of some lament, a "bifurcation" of nature.

If Whitehead thought of having a personal mission, it was to rid philosophy of the concept of substance. For this purpose he stripped from his vocabulary every word which might suggest the culpable idea. Things became events; in their individuality they were "occasions." Their intimacy and movement he designated by the word "concrescence," to show that things "grow together" like seed and soil.

If you lived upon a great expanse of ice and one day saw the bergs dissolve and a warm blue ocean appear, the transformation would not seem more remarkable than that produced by Whitehead's "philosophy of organism." To come hither from the Hegelians (who had everything nailed down), from the pragmatists (who had everything torn up), from the realists (who had everything dried but not salted), was to suffer extreme astonishment. At first you thought that the chief difficulty was the language; then you discovered that the ideas themselves were strange. Your training had not prepared for them.

But it was possible to learn. I'm not sure that I could have reached, or have thought that I reached, the doctrines of *Process and Reality* without Whitehead's expositions in seminar. I needed to turn the notions over many times before clarity might set in. And in these labors I was not a little distracted by the many epigrams — some of them pregnant and all of them witty — which showered from his discourse.

You may fancy the scene: the small man with the great head and smiling "cherubinnes face" over the pointed collar, pouring out a cup of tea from his Thermos bottle (it being four o'clock) and settling thus into an exploration of the universe; we students sitting around the table, each with a copy of *Process and Reality* and the inevitable earnest notebook (mine, I was alarmed to discover, contained nothing but the epigrams). And the voice, charming in itself but made more so by the lingering of a childhood stammer: the *r*'s softened almost out of being, and every consonant made gently liquid.

And then the exploration itself: what Bradley had seen and not seen, what treasures lay forgotten in Locke, what needless skepticism there was in Hume, what baffling naïveté in the positivists. "There ought to be a separate department of Kant, to remove him from modern philosophy." "When Hegel found a contradiction in his own thought, he mistook it for a crisis in the universe." "Russell is always telling us what we can never know."

That was my introduction to dialectics. The path was perhaps not straight, and when one came to the end of Whitehead's part of it one still had some distance to go before one came to Marx. What one could not do, however, was to turn back. Once the concept dawned of a universe immaculately whole, a universe of order *and* change, of sameness *and* novelty, of preservation *and* achievement, who could have given it up? To do so would have palsied theory and lamed practice.

That I have followed the path to its end these pages must make plain. Given the economic crisis and the absurdity of official explanations, I turned to the writings of the Marxists, which, if they are the last to be read, are also the last to be forgotten. The book fell open, by a kind of fatality, at a passage in *Anti-Dühring* in which Engels discussed the very phenomenon then before my eyes:

Their political and intellectual bankruptcy is hardly still a secret to the bourgeoisie themselves, and their economic bankruptcy recurs regularly every ten years. In each crisis society is smothered under the weight of its own productive forces and products of which it can make no use, and stands helpless in face of the absurd contradiction that the producers have nothing to consume because there are no consumers.

I could not then conceive, nor can I now, what alteration in this passage truth might possibly require.

So my journey is taken, and I am content. Yet I would not urge it upon anyone unless he is sure that he will find the destination joyful and of good fame. For there is no road back, but only a fall, a coward's fall, into gulfs beneath contempt. The case is just as Milton demonstrated: hell is full of comfort and disaster.

For those, however, who have learned to live at the end of the road there is a vision always accomplishing itself, a vision of human perfection rising out of history and of history rising toward that goal. Faith (if, in the end, we must call it that) is the assumption that the universe makes this rise possible, hope is the expectation that we can rise thither by our own powers, and love is the knowledge that we must make the ascent together.

It may be that in the future there lies some theory more adequate to the purpose than this; but, as things stand thus far, I have not found its equal among the philosophies.

KNOW JUST WHY

CHAPTER VI THE KNOW-
NOTHINGS AND
THE KNOW-IT-ALLS

Having now through three chapters spoken in large about the cosmos, we must expect the question, "How does one *know* that all this is the case?" Doubtless the question would have been asked already, if books permitted conversation between writer and reader. As things are, however, the reader's power over a book lies in the fact that he can shut it. The writer's power lies in the fact that he can speak right on without interruption.

The question is legitimate, and will be perfectly honest provided one grants that the theory of knowledge involves a theory of the universe quite as much as the theory of the universe involves a theory of knowledge, and that the two theories have to be developed in conjunction with each other. Any attempt to describe the universe raises questions about the method of describing it, and at the same time the success of any method can be determined only by what we take the universe to be.

The problem of knowledge, legitimate as it is, can

nevertheless be abused. One of the classic devices of controversy is to shrink discussion from ends into means, from consequences into methods, from things known into methods of knowing. There are whole philosophies of just this sort, and they are capable of devastating effect.

The connection between human knowledge and human control over events makes the theory of knowledge the very nerve of action. That nerve is readily paralyzed. If, for example, people seriously believe that knowledge is impossible, they lose the hope of controlling events and are left forever undecided what to do. Or if (it is the contrary fault) they are "too sure of themselves" by virtue of a certainty not scientifically grounded, they risk error in the incorrigible form of prejudice; and, since error entails the lack of control, a long series of defeats will lead these people to failure and even tragedy.

Apparently it is bad for men to think either that they know nothing or that they "know it all." I should imagine that this is a very simple fact which most of us learned early in our lives. At any rate, it is supposed to be one of the fruits of education. Nevertheless, it is so far from being universally recognized, that there are philosophies of considerable power maintaining each of the extremes. I think that these philosophies don't appear in all times and circumstances, but they have a way of turning up, like gods out of the machine, to rescue rulers whom science has placed in difficulty.

The view that there is no genuine knowledge because all opinions are about equally cogent may be called, from popular parlance, "relativism." The other extreme, which holds that men have a nonscientific basis for absolute certainty, is constituted of two doctrines: (1) that there exists some personal authority which can render statements infallibly true, and (2) that there exists a technique of private mystical insight which can do the same thing. All these views are interesting in the way that diseases are interesting, and the present chapter undertakes a study of them as part of the pathology of the human mind.

The Social Basis of Relativism

In my boyhood I used to know a lady whose method of debate was to answer every argument with the words, "That's only your opinion." A great deal hung upon the word "only," one of the most treacherous in the language. For obviously the argument was at least an opinion, and the opinion was at least the arguer's. The power of the remark lay in that "only," which suggested that the opinion was not necessarily true and not necessarily held by anybody else.

The remark itself had a kind of beauty, which was the beauty of a circular argument. For, in order to know that the opinion was nothing more than an opinion (i.e., not an item of true knowledge), one would have to show that the opinion was at least doubtful. This being, however, the point at issue, the lady had advanced to her conclusion without passing through any proof. The absence of proof was, no doubt, partly amended by the plunge and irony of the utterance.

Now, if such a statement is raised above the level of mere retort and is turned into a theory of knowledge, it becomes in fact a theory of no-knowledge. The more orthodox and certainly the more erudite name for it would be skepticism or Pyrrhonism (after Pyrrho, a Greek, who held this doctrine most skeptically). But the form in which I hear it commonly stated is the expression "Everything is relative," which sentence apparently means that everything is relative to one's opinion of it.

Consequently "relativism" is the name I prefer and shall use, dissociating it from anything so lordly as relativity theory or the logic of relations. One cannot wish to blur in any way the distinction between what is the total defeat of knowledge and what is the triumph of it.

One road to relativism issues from the sheer difficulty of thinking scientifically and from the discouragements which attend that effort. We observe that everybody has opinions. We find it hard to prove other people's opinions wrong, which means (by logical equivalence) that we find it hard to prove our own opinions right. It may seem that all faces can be saved by letting opinions drift as they may, each no better than another.

Another road issues from what might be called the perplexity of princes. For there are in the world certain ruling groups who attempt to conduct the world's affairs singlehandedly, forgetting that this sort of thing can be done only by mankind as a whole. When men of power find their hopes abortive, they renounce the knowledge on which those hopes were fed, and discourse through their ministers of grace, the intelligentsia, many consoling words on the ubiquity of doubt and the frailty of conjecture. If a man sets out to rule the world, and ends by having the world rule him, the least he will feel is that life is very uncertain.

Before this melancholy change completes itself, a moment occurs when the mighty man, seeing his power frustrated, sees nevertheless that he must keep it or vanish out of time. Defeat has weakened confidence in his own knowledge, and has suggested that there is other more potent knowledge which he has not. What if this knowledge belonged to other men? Rather than this, let doubt prevail and ignorance be universal.

Thus, however real the perplexity which feeds this doctrine, it has a convenience deeply stained with guile. If, for example, we wish to examine a piece of reportage for distortion or even for lies, our only means is to set the narrative beside the facts. This test, which is decisive though difficult, is of course the one which distorted narratives try frantically to flee. For this purpose they will usually limit misstatements to subjects on which information is hard to get or easy to hide. But obviously they will fare most safely on the assumption that there are no objective facts at all, and therefore nothing to which the narratives are answerable.

Many of us have long supposed that this must be the secret belief of journalists, but I will confess to some astonishment at finding so frank a statement, editorially given, in the *Washington Post:*

> There is no objective standard for determining what is news any more than there is such a thing as objective fact. Facts appear to be facts only because a majority of the people accepts them as such.[1]

[1] Issue of November 23, 1948: editorial, "What Is News?"

The first sentence brings the heavens down. The second erects an umbrella to fend off the tumbling planets.

Thus the motives which lead men to relativism are sometimes perplexity, sometimes weariness, and sometimes deceit. None of these gives much dignity to the doctrine, and we could ignore it altogether if it were not for its profound social effects. The world, as Sean O'Casey's Captain Boyle used to say, "is in a terrible state o' chassis." Part of the crisis is a strife of opinions. During the course of struggle some opinions prevail and others languish. Now, if it is held that opinions are indistinguishable as regards truth and falsity, the whole problem of making true opinions prevail simply does not exist. There is no longer any hope of rational guidance over human affairs.

Some refutation of the theory seems therefore required. To achieve that refutation fully, we must, I suppose, describe the conditions under which knowledge is possible. However, erecting an entire theory of knowledge for the sole purpose of destroying relativism seems like exerting an elephant's strength upon the obliteration of a gnat. Our present need will be satisfied if we can show that relativism, whatever its charm, is nevertheless incredible. This will not prevent the gnat from stinging, but it will show that the stings are only the stings of a gnat.

The Doubter Doubted

We must now try to get a precise formulation of the doctrine we are disputing, especially because it presents itself, in very favorable guise, as a deduction from probability theory. Such a deduction it is not, and we must strip away this flattering parentage.

Everybody is aware that the assertions we make about the world differ in respect of the chances they have of being true. For example, it is extremely probable that the sun will rise tomorrow, and it is extremely improbable that Santa Claus exists. There are degrees in between: it is probable that you will live to be fifty years old, and it is possible that you will live to be ninety.

Furthermore, everybody is used to acting upon probability, to considering probability (in Bishop Butler's words) as "the very guide of life." Suppose you are a Philadelphian and decide (as Philadelphians sometimes do) to go to New York. Your decision to go and the behavior consequent upon it are based upon the expectation of getting there, despite the "off chance" that you may die on the way or the train may break down or the train may travel in the opposite direction. You don't know with certainty that you will get to New York — in fact, you know what some of the possibilities are that you won't — but there is sufficient probability in the expectation for you to travel with serenity.

Indeed, it is very possible to act upon an improbability. You ring a friend's doorbell "on the chance" that he is in. And in great affairs, where you are following a policy of the sort pompously called "calculated risk," you will keep alert for signs of the improbable negative case which you have discounted.

Reliance on probability doesn't undermine confidence in our knowledge and in no way paralyzes action. Evidently this is not relativism, which *does* undermine confidence in our knowledge and does paralyze action.

Nor can any view which does these things be inferred from probability theory.

Relativism, then, doesn't say (as its own peculiar doctrine) that there are no assertions which we can with certainty know to be true, and that we must therefore be content with probable knowledge. It says, rather, that there aren't even any degrees of probability, that *all* assertions occupy one dead level of validity (or, if you like, of invalidity) regardless of the evidence gathered for them and regardless of their logical consistency.

The relativist doctrine can, accordingly, be stated thus: *There is no one sentence which has a greater probability of being true than any other sentence.*

Now, this statement of the relativist doctrine is, of course, itself a sentence. If what it says is to be taken as applying to itself, then it has no special probability of being true. It defeats itself, as Pyrrho did when he doubted his own skepticism.

However, self-referent sentences are often in a pickle logically; and, in asserting one of these, relativists wouldn't be worse off than many another philosopher. We can restate the doctrine in such a way as to exclude self-reference, thus: *There is one and only one sentence which has a greater probability of being true than any other sentence, namely, this sentence itself.* Or we could say in more colloquial language: One thing only we know, that we don't know anything else.

What grounds can there be for our accepting such an idea? They can't be rational grounds, for then the relativist assertion would be inferrible from other assertions, but these (according to relativism itself) have no probability of being true. Relativism is in no position to call on logic or scientific method.

Nor can relativism base itself upon authority. That would involve other sentences naming the authority and demonstrating its authoritativeness; and these, being *other* sentences, also lack probability. The only ground I can imagine on which relativism might be accepted is direct, quasi-mystical insight. No other sentences would be involved in that. Once in a lifetime the flame would come, and out of the flame a voice, crying:

> Nothing is truth, truth nothing — that is all
> Ye know on earth, and all ye need to know.

I don't suppose such grounds are very persuasive, and I have an idea that the propounders of relativism maintain the doctrine by not letting themselves see exactly what it is. A little skin will cover dislocated joints; a little fuzziness will hide absurdity. With this salutary thought in mind, we shall return for a moment to our friend the Washington editor, who was in the unfortunate case of being a thinker without being a logician. He wrote, as you will remember:

There is no objective standard for determining what is news any more than there is such a thing as objective fact. Facts appear to be facts only because a majority of the people accepts them as such.

There is argumentation in this passage, though the joints of it are not clear. The course of the argument I take to be the following, and you must decide for yourselves whether I put it fairly:

If there is to be an objective standard for determining what is news, then there must be such a thing as objective fact. But there is no such thing as objective fact ("facts" being merely majority opinion). Therefore, there is no objective standard for determining what is news.

This is the pattern, familiar to logicians, of *Modus Tollens:* given the fact that one statement implies another, and given the fact that the implied statement is false, we infer that the implying statement is false too. "If wishes were horses, beggars would ride; but beggars don't ride, and so wishes are not horses."

What is singular about the editor's argument is that he holds the two premises, the conclusion, and the appropriate rule of inference to be facts quite independent of majority opinion, thus:

Fact No. 1: If there is to be an objective standard for determining what is news, there must be such a thing as objective fact.
Fact No. 2: There is no such thing as objective fact.

From these he infers a third:

Fact No. 3: There is no objective standard for determining what is news.

And in making this inference he assumes:

Fact No. 4: That Fact No. 1 and Fact No. 2 imply Fact No. 3.

By using the Lewis Carroll paradox, which shows that in every inference there is an infinite number of assumptions, we could provide the editor with enough facts to occupy him throughout time and eternity.

An argument purporting to involve four facts can hardly contain a minor premise which says that there are no facts at all, for either that premise will have to be false (i.e., not a fact) or the argument will be merely fanciful. But if the minor premise is false, then the conclusion becomes doubtful, and it may then be the case that there *is* an objective standard for news after all. In the meantime the four "facts" have melted under fire and have shrunk to two. It is better for relativists to let argumentation alone.

At the risk of seeming relentless, we ought to hunt the quarry a little further. The editor asserts that there is a difference between *objective* fact and what a majority of people accepts as fact. The first of these, he says, does not exist. This will be because there are no events occurring independent of observers, or because the observers so modify events in apprehending them that subsequent reporting shows the state of the observer's mind rather than the state of things.

But what of these alternatives? Are they facts, truly objective and occurring independently of observation? Or are they the fatally biased visions of limited observers? If they are the first, the editor is self-refuted. If they are the second, we have no reason to think them true.

Such are the agonies of relativism trying to make itself credible. Of all philosophies it has the feeblest intellectual supports, because it is the negation of everything intellectual. Rather, it is a fashionable frailty, a broth of indecision, an ambient and aimless air, congenial no doubt to men who like such modes, such tastes, such weathers. But I think no one would hold it who wished to leave upon history the mark and record of a useful mind.

From Know-Nothing to Know-It-All

Early in each semester there comes the moment when my students put (as it seems) the unanswerable question, "Who is to say?" By this time we have advanced into the tangles of philosophy, the thread of argument has disappeared, my further discourse seems like a proclamation, and so the students politely suggest that, though there may be authorities who can settle the question, I am not necessarily one of them.

"Who is to say?" The question is partly torrid and partly horrid. The heat comes from anarchist fuel: students may mean that they don't want anybody to tell *them*. But there is also in the question — and this is the horrid part — an implication that if the truth of a statement were to be determined at all, it would have to be determined by personal authority. So ridden are we by "experts," by the prestige of men who live in print, that the youthful mind, if it bows at all, bows to authority rather than to proof. Only by a judicious application of the laws of shock can I convince students that it is not a man's authority which makes his statements true, but rather the truth of his statements which makes a man an authority.

Thus, among the many things I have learned from teaching (and these are more numerous, I dare say, than the things I have taught), one of the most striking is the connection between relativism and its apparent opposite, authoritarianism. But when you think about it a little, there is not much difficulty in perceiving why the connection exists.

Relativism is discontent with rational criteria. Logic has been examined, and dismissed as unconvincing. Science has been found infected with doubt. Moral principles are seen to be fantastically feeble, since in a world where we are ignorant of everything except our own ignorance we must also be ignorant of right and wrong.

The result of this, for anyone who has to live and move upon earth, is catastrophic. Suppose that rational criteria are the only possible criteria, and suppose that these do not exist. Then there are no criteria whatever. Consequently, we cannot know what situation confronts us at any given moment, what alternatives there are, or indeed whether there are any persons answering to the pronoun "we" and confronted by situations and alternatives. Nor can we know what choices we ought to make, either out of selfishness or altruism.

Such a conception is intolerable: you cannot seriously entertain a doctrine you cannot live by. But if, realizing this, you still doubt science and reason, then you will turn to standards of another sort. These will have the special character of claiming to decide all issues without using any of the means which science would employ.

For example, you won't find in any scientific demonstration an assertion to the effect that such and such a statement is true because Newton or Darwin or Einstein says it is. On the contrary, you will find a belief that anything said by Newton or Darwin or Einstein is as open to criticism and analysis as anything said by anybody else. There will be, no doubt, a respect for great minds because of their achievements, but these achievements will never be regarded as final for the human race.

Nor will you find in scientific demonstration any belief that the private feelings of the investigator are what makes the proof convincing. No physicist would say, "The sentence 'For every action there is an equal and opposite reaction' gives me intense joy and is therefore true." Neither would he say, "I know a great deal about the world, but I cannot communicate one particle of that knowledge."

Evidently, then, scientists don't base their reasoning on personal authority or on private unutterable feelings. These two, however, are precisely the criteria to which a distrust of science must lead. A man weary of doubt and weary of thinking flings himself at last upon the rock (as it seems to him) of perpetual certainty. The know-nothing becomes the know-it-all.

Thus, in a curious way, a relativist is a man of inverted omniscience. Turn him right side up, and you have an authoritarian or a mystic. The new posture brings him no nearer to knowledge. Instead, it supplies the world with further erroneous views. These we must now examine.

Ipse Dixit

Authoritarianism is the view that truth is determined, not by principles, but by persons. Even in its supernatural form, this is what the theory asserts. For suppose you say that the only absolutely true statements are those which have been divinely pronounced; evidently you are regarding God as the sort of being who can make pronouncements — that is to say, as a person. In fact, according to the Athanasian Creed, God is redundantly personal, since he is three persons in one.

The same condition holds for "infallible" documents, such as creedal statements, conciliar edicts, and the Bible itself. If these are taken simply as the work of men, they will be open to the same critical and historical analysis as other works of antiquity — the Homeric poems, for example. But if they are taken as the work of men inspired, then the authority of inspiration suspends criticism and demands assent. The inerrancy of the document derives from the perhaps temporary infallibility of the authors, who hold it from the Triune Person Himself.

Apparently this is true, also, for admission, after a span of years, into Heaven. In recent months admission has grown more difficult than once it was, and may be canceled by the simple reading of a printed page. A man of left-wing sentiments can now enter Heaven no more readily than he can enter the United States, admission to both being determined by consultation between the hierarchy and the Attorney General of each commonwealth. Thus the predictable truth of the statement "Heaven's my destination" depends upon the authority of several persons, of which the human and mortal have recurring intimacies with omniscience.

A long tradition has likewise preserved this point of view in ethics. The early lawgivers were careful to present themselves as recipients of revelation from on high. You had to climb a mountain before you could legislate.

You thus gained eminence and solitude, and your transactions with eternity were unseen.

From this procedure it is difficult not to infer a lively skepticism among ancient peoples. They knew their lawgivers too well to suppose them capable of legislating without supernatural aid. The habit thus set has come down to us. A series of theoretical refinements have organized ethics into a single body of law pronounced, at the beginning of things, by a single deity.

The skepticism apparent in ancient peoples is by no means dead, and it still produces its characteristic effect, that of driving authority up to the supernatural. Experience of error is so frequent and so universal that only an astounding credulity would enable us to believe any man, so far as he is a *man*, infallibly right. Hence it is safer and less fertile in doubt to move the infallibility upwards and let selected mortals acquire it by descent. Among the several benefits of this view there is this, that infallibility need not rain down continuously, but may be left to drip at judicious intervals.

Such people in such circumstances are fortunate. With laymen, who cannot be suspected of these supernatural affinities, the case is very different. For how can they persuade us of even a sporadic infallibility? Length of study, length of years, length of beard may suggest it; membership in learned societies may give color to it; a man may even reach that peak of fame where he is accurately quoted in the journals. Nevertheless, suspicion will infect the human clay, and no man will, by sheer vastness of reputation, be thought infallible.

So much for the hazards of personal authority. The question chiefly to be answered is whether we get any knowledge from the use of such a standard. Is it really the case that certain statements are true because some authority, human or divine, has affirmed them to be so? And has the authority also been so generous as to establish the rules of inference by which we can draw conclusions from the immortal premises?

I suppose we shall first be curious to know how we can recognize the authority as authoritative. He cannot be *proved* to be so, for this would require the truth of other assertions. These other assertions would have to be true either independent of the authority (in which case the authority is not ultimate) or dependent upon the authority (in which case the argument is circular). By rational criteria, then, the argument for authoritarianism must be either incomplete or fallacious.

Accordingly, we reach the same result as with relativism. There we had to intuit a statement which presented itself as the one probably true statement. Here we have to intuit an authority who presents himself as the only valid authority. Now, where proof is impossible and intuition is necessary, there has to be something in the statement which can induce rational belief — the sort of thing which used to be called self-evidence. Or, at the very least, the statement ought to commend itself to us as a "working hypothesis." But neither of these attributes belongs at all to relativism or authoritarianism. Consequently, in both these views appears the character once held to constitute the nature of a gentleman: they have no visible means of support.

I suppose, however, that the power and function of authoritarianism never did derive from its being rational. Rather, they derive from its being coercive.

Beneath an aspect of noble principle there lurks the threat of real force, chastening the unbeliever. Losses terrestrial and celestial are to be incurred, pains physical and spiritual are to be suffered, unless the authority and its pronouncements are duly acknowledged. Because of this, great numbers of men, out of simple immediate need, abandon rational standards (with which, perhaps, they were never allowed acquaintanceship), and decide to believe what they are told.

These considerations suggest what is, I think, historically true: that authority, once plainly fixed in person or institution or document, turns out to be that of a social class possessing actual power. The authority of the medieval church upon matters of dogma was in fact the political power of a dominant section of feudal aristocracy. This was replaced, after the Reformation, by the authority of inerrant Scripture, which represented in its turn the power of the Protestant middle class. All authoritarianism in thought is class-power in fact.

The persistence of authoritarianism is partly due to government as such, and anarchists are quite right when they point this out. To command is the very nature of government; its pronouncements may not be true, but they require acceptance. For most citizens there occurs at length a blurring of the borderline between science and politics, and truth begins to associate itself with the strongest pressures. It takes a very firm nature to endure an interminable divorce between true statements and safe statements. One feels much more comfortable thinking that the safe statements are true.

Much of this confusion is avoidable, and one can in fact measure the integrity of governments by their willingness to permit clarification. Suppose, for example, a government were to say, "These appear to be the facts, and these will probably be the results of our policy." Then, in enforcing the policy, that government would have drawn a clear line between authoritarianism in politics and rationality in science. It would have humbled itself before the standard of truth, while at the same time exercising its mastery over the citizens.

But unfortunately when governments get excited — and they are the most excitable form of protoplasm on earth — the rationality vanishes even as the authority grows. Absolutes rain down like hail, shattering the common, peaceful windowpane. Scientific probability gives place to that most unsure of certainties, fanatical conviction. But behind the icy curtain, the great forces of history move on, unseen and uncontrolled. In the end, the hailstorm blinds the men who raised it, and leaves them prostrate beneath the powers they had thought to hide.

One may observe, if one likes, the self-destroying process. For example, there was recently a time in our government when the proposition "Organization X is subversive" rested not on proof (for no proof was permitted) but on the authority of the Attorney General. Now, proof would have contained, as one of its necessary parts, a criterion of subversiveness valid for all possible cases. This being absent because suppressed, could any given Attorney General reasonably object if his authority were superseded by some other authority which thereupon pronounced him and his entire party subversive? If he then resorted to argument, we should wonder why he had not accepted argument in the first place. If he did not resort to argument, he had simply to acknowledge the greater power of the new authority.

It is one of the idiocies of centrist politics that the center undertakes to be authoritarian toward the left and rational toward the right. The grievous end of this is, of course, the victory of fascism, which is authoritarian toward everybody. If the content of a list depends on nothing but the will of the compiler, the ultimately authoritative list will be made by those who have, not the greatest reason, but the greatest power. Consequently the Senate "liberals" who conceived and enacted concentration camps may be the first persons to inhabit them.

Politics, indeed, is a perpetual demonstration of authoritarian follies. As we observe each witless act nearing its inevitable doom, we can see more and more clearly why authoritarianism cannot possibly be a standard of truth. For, taken as standard, it distracts our attention from the actual course of events to the thoughts of some person or group of persons. Be these as authoritative as they may, the process of events remains just what it is, and the facts are what they are. A man, by twisting your arm, may make you say, "What lovely sunshine!" when in fact there is a torrential rain; but he does not abate the rainfall by a single drop.

If human knowledge aims at recording what actually goes on in the world, it will have to consult the world rather than this or that person. It will admit authority only in the form of "expert opinion" — the opinions, that is to say, of persons whom we can reasonably suppose to have an intimate and detailed acquaintance with a particular set of facts. But the authority of such persons comes from their scientific stature and not at all from their possessing political power. In other words, it derives from a standard which is valid for all thinkers, including themselves. They do not know because they are authorities; they are authorities because they know. In the end it is they who must triumph, and we with them, so far as we can do likewise.

The Superluminous Gloom

In the world of our political experience, authority over thought and power over people are the same. Rather too obviously so, indeed. From this follows, as we saw, the tactical device of lifting authority into the clouds, where it may utter oracles.

Acceptance of an authority thus elevated requires a rather special posture of mind. Other realms, other methods: if the fount of authority is transcendental, the path to it will be transcendental also. Authoritarianism, escaping the imperfections of humanity, allies itself with mysticism; and the obedient spirit, swift at its master's summons, rises to explore a darkness beyond day.

This darkness was known to Dionysius the Areopagite as "the superluminous gloom." I have the impression that, whatever may be said about the adjective, the noun is accurate enough. Yet, apparently, for spirits capable of the necessary flight, the gloom is streaked with flashes and in each flash some newer universe stands revealed. Then the same spirit which soared among the lightnings drops down again to the low and level world, the common day.

Whether or not these flights teach wisdom, they do,

it seems, teach liberty. The human spirit, moving so easily through air and mist, learns to love its visions — first, because they are visions, and then because they are its own. It has abandoned science long ago; shall it not abandon authority as well? Faith, which gave it wings, becomes the assurance of truth; and faith is feeling — inward, secret, incommunicable feeling — which owns no master but itself.

For this reason the churches have looked on mystics with some suspicion, and the mystics have not infrequently denounced the churches. Indeed, how can a church tolerate spirits who, after visiting the upper air, report that things are not what dogma says they are? On the other hand, what is the use of visiting the upper air if its tracts and spaces are already drawn? Authority needs the supernatural as a hunted man needs a cave. But the supernatural needs authority no more than a cave needs a man: it will be a little fuller with him, but, without him, it will still be a cave.

The bright messages of mystics have all a common tone. They tell us that the truth of a sentence is measured by the intensity of feeling to which the sentence gives rise. Gives rise, that is to say, within the mystic, for clearly he cannot be concerned with other people's feelings, which he does not have and cannot directly know. By such a standard the recognition of truth is entirely private and personal. The mystic is saying, "I have before my mind a certain sentence s, and s, thus entertained, arouses my feelings in a high degree. The degree, moreover, is high enough for me to recognize s as true."

In fairness one must say that mystics don't apply this

THE KNOW-NOTHINGS

test to commonplace sentences. For example, they wouldn't talk in this way about "My toothbrush has a green handle," for it seems unlikely that such a sentence could be a lure for passion. Mystics would probably consider sentences like this to be a sign of the dreadful triviality of sense experience, and would leave discussion of them to the more pedestrian philosophers.

The mystical standard is therefore reserved for "crucial" statements — those on which human happiness may seem to depend. Such a one might, for example, be "My soul is immortal," or, to put it perhaps more accurately, "I am a being whose consciousness will not be ended by physical death." It falls, moreover, outside the reach of empirical confirmation (the "bourne from which no traveler returns"), and thus invites the use of some other standard than the one which would verify "My toothbrush has a green handle."

Given, then, a group of assertions which seem crucial and which neither logic nor science is able to verify, we have the characteristic themes of mystical insight. And indeed, if the mystics are read as poetry, there will be no harm. The story of Saint Francis converting the wolf of Gubbio, if it were taken literally, would be a vulgar, preposterous tale. But if we understand its fictions to mean that one man can do more by love than many men by slaughter, we have a sentence verifiable on the usual scientific grounds.

Unfortunately, mystics do not stay within their proper limits, but pass over into politics and sociology, where they begin "intuiting" such things as the inferiority of Negroes and the sublimity of industrialists. These excursions, which may become habitual, are the reason why we must trouble ourselves to analyze mysticism. Let us say, first of all, that emotions do signify something. What they signify, however, is not, except indirectly, a state of affairs outside the body which has them. Saint Catherine of Siena, for example, had a vision in which the infant Jesus put a wedding ring upon her finger and made her his bride. This tells us nothing (except inferentially) of events going on outside her body, but a great deal of what was going on within it.

Or suppose we have a man who ardently believes that someone is trying to harm him, when in fact no one is. His belief tells us nothing about other persons, but something very important about himself, namely, that he is paranoid. The very intensity of his conviction suggests that it has subjective origins rather than objective ones.

Matters like these are commonplaces of psychiatry, which is probably the science most devastating to mystics. The existence and intensity of a certain feeling in a certain person do indicate the condition of that particular brain and nervous system, though even here we shall need more facts before we are fully enlightened. It is also true that the existence and intensity of the emotion will indicate that there are things outside the body affecting the body in a certain way. But unless both the emotion and its intensity are interpreted from the point of view of an analyst, they contain nothing which will show precisely what the actual environment is, or whether the belief so excitedly entertained is true.

Tennyson drew his wounded spirit through a long but efficacious cure, and in the end sorrow made him a

THE KNOW-NOTHINGS

laureate. But while the therapy was proceeding and the dismal clank of pre-Darwinian doctrine smote his helpless ear, there came the moment of defiance, the ecstasy of reckless assertion, the flourish of the mystical flag:

> A warmth within the breast would melt
> The freezing reason's colder part,
> And like a man in wrath the heart
> Stood up and answer'd "I have felt."

This is the wrath which succeeds upon the collapse of rational argument, and is a kind of resolute testimony that the believer will continue to believe despite the lack of evidence, or even, perhaps, despite the presence of it. The information it thus conveys has to do with the believer and not with the object of his belief. We know that Tennyson very much wanted immortality to be true, and we can guess from his frequent recurrings to the subject that he always found it rather hard to believe. But his feelings, however poignant, do not in the least demonstrate that immortality is a fact.

The second difficulty with the mystical criterion is that it makes truth private and personal instead of public and social. Now, scientists expect that they can verify one another's results; and philosophers, though somewhat more individualistic, suppose that their arguments, if valid, will withstand criticism. But nothing of this sort can touch a mystic. It does not matter if *his* statements are contrary to logic or to sense experience: all he ever required of them was that they be profoundly felt. If William Blake chooses to believe that a thistle by the roadside is really an old man with a gray beard, there is nothing you can do about it in a philosophical way. He has his own "truth," and no more may be said.

It might seem, when you first consider it, that if truth is a matter of intense personal feeling, an absolute chaos will reign over human opinion.

> Thy hand, great Anarch, lets the curtain fall,
> And universal darkness buries all.

This is a formal possibility, but not, I think, the probable social fact. The opinions most intensely felt are those connected with profound personal needs as these are satisfied or frustrated in society. There seems, indeed, to be some relation between the thwarting of desires and the intensity with which certain opinions are felt. A satisfied, well-adjusted person entertains ideas more coolly. He doesn't feel that "life will lose all its meaning" if any of his ideas prove false.

We have therefore some reason to think that the most intensely felt opinions reflect the insecurity of men in their relations to one another, the defeat of cherished hopes, the guilt of committed wrongs. It is on account of this that exploiters must have a sort of ritual mystery which will affirm the baseness of the people they exploit, and the exploited tend to believe fantastic notions as to who their persecutors are.

The effect of mystical criteria, then, is to unstop the sources of error, particularly on those subjects where, our desires being balked, we most need to be rational. A burdened life doubtless suggests the hope of a happier eternity; but this hope, though it falsely lightens the heart, will not lift one faggot from bent and weary backs. To the extent that mysticism advances, humanity retreats. The power of mystical criteria is equal to the helplessness of man.

THE KNOW-NOTHINGS

Where knowledge is concerned, then, there is no hope in personal authority, however sublime, nor in personal feeling, however intense. The know-it-alls are as feeble as the know-nothings. Both groups, we may perhaps say, are a nuisance developed under social pressure into a menace. Verification by fiat or by feeling is no verification at all. Consequently relativists, authoritarians, and mystics are, in effect, in a state of general ignorance, spotted, it may be, by random knowledge.

If such a state seems comfortable, I should think it can seem so only to men who do not know they are in it. Ignorance no doubt makes ignorance tolerable. But people who need to know a certain thing and are aware that they lack the knowledge, will never content themselves so easily. They will seek, rather, the standard and the methods of knowing, and, having these, will turn to the problem in hand.

It is time for us, after dallying with doubt, submission, and ecstasy, to do likewise.

CHAPTER VII QUEEN TRUTH AND
KING CHARLES

Mrs. Nixon has probably never seen a relativist, though mystics and authoritarians, I suppose, are known to her. It is interesting to imagine what would happen if she, meeting each of the three, put to them her pregnant question, while they made their formal philosophic replies.

"I wish," says Mrs. Nixon, "I could explain just why and know just why we have such a hard time in this part of the country."

"I, too, wish that that were possible," says the skeptical editor. "Unfortunately there is no such thing as objective fact, and if that's what you're trying to know, you will simply have to give up the effort."

"But," Mrs. Nixon replies, "I have six children, who get hungry and cold. I'm not just imagining them. And there's my husband, who was shot. I didn't just imagine that."

"It seems terrible," says the editor, "but really there is no objective standard for being certain of these things. Facts are facts only because a majority of the people accepts them as such."

"Wouldn't most people think I have children and that my husband was shot?"

"Possibly. It depends on the newspapers."

"But won't the newspapers say what happened?"

"They will say what a majority of the reporters think happened."

"And what will the reporters think, then?"

"They will think whatever their editors regard as advisable to be thought."

"Then I might be imagining all this trouble?"

"You might and you might not. How can one know? Most things seem bad only because we look at them that way. If you would, ah, shift your mind a little —"

"It's strange, very strange. I sit and think of him when night comes. And your way of thinking isn't mine at all, not at all."

There is a third voice and a third face, both grave, both commanding. "The Bible says, *The poor ye have always with you.*"

"Does the Bible say that I am to be poor?"

"It also says, *Man doth not live by bread alone.*"

"Does it say that men can live without bread?"

"Economists say that food is more plentiful when men make a profit by selling it."

"If I had money, I'd be glad to buy."

"Biologists say that if you don't have the right genes, you can never have much money."

"Did they ever examine mine?"

"Political scientists say that if you have a vote, you are in a democracy, and everything is all right."

"And I say that if you have a vote and don't get shot for using it, then everything is much the same."

"These are wise men, Mrs. Nixon. You must trust them and follow them."

"I wish I could explain just why and know just why wise men seldom tell me anything that is helpful."

There is the fourth voice and the fourth face, both bright, both passionate. "Trust to your inward self, my child. Forget the outer world, which is illusion, and sink deep into the true reality, which is You."

"But I don't find anything there but trouble."

"That is because you don't yet know yourself. Kierkegaard said, *Subjectivity is truth, subjectivity is reality.*" [1]

"Who was Kierkegaard, and what did he mean by that?"

"He was a great Dane, and he meant that the only real knowledge you can have is to know yourself."

"Did he like knowing himself, or did he wish he could know somebody else?"

"He liked it very well, for he perceived that, though spotted and even sinful, he was an immortal soul. But he could not wish to know other people, because he could not know them except as a possibility."

"You mean that I can't *know* that there are children of mine, or that they are hungry? It's merely, perhaps they are there, and perhaps they are hungry?"

"Yes. You see, *with respect to every reality external to myself, I can get hold of it only through thinking it. In order to get hold of it really, I should have to make myself into the other, the acting individual, and make the foreign reality my own reality, which is impossible.*" [2]

[1] *A Kierkegaard Anthology,* edited by Robert Bretall, Princeton University Press, 1946, p. 231.   [2] *Ibid.,* pp. 226–227.

"How am I to feed my children unless I know that they are there and need food; also unless I know where food is and how I may get it?"

"For the mere outer world, possibility is enough. Come, instead, into the inner world, where feeling radiates, and every beam lights some new truth!"

"Can I do all this without eating?"

"No, my child, you will have to eat something."

"Then, in order to enter the inner life, where the sure things are, I have to eat some outer food, which is rather uncertain?"

"Well, after all, you needn't eat much."

"Sir, do you know what I think? I think you mean that for people like you eating is a certainty, and for people like me eating is just a possibility."

There could be other voices and other faces, which would be ours. But unless we come to Mrs. Nixon with a rational standard of truth and with methods for applying it scientifically, we shall be of no more help than her previous visitors. If all we offer is comfortless exhortation, she might as profitably converse with the fowls and animals of the farm.

Society and the Standard

The three theories we have thus far talked of agree in respect of the fact that they all make truth somehow personal and capricious. They assert, respectively, that truth is (1) anybody's opinion, (2) somebody's edict, (3) anybody's passion. Relativism has the equality of indifference, mysticism that of anarchy. Neither of these has any rule which could require all thinkers to think alike so far as they think truly. Thus the law of contradiction is suspended, and assertions can be made without the slightest risk of denial.

It might appear that authoritarianism escapes this difficulty because the authority's statements are held to be binding on everybody else. The trouble is that the authority is not subject to the same rule which he imposes on others. He decides personally and capriciously. He presides over a hierarchical system, with an upper class of truth-seers and a great lower class of obedient ignoramuses.

The failure of these three theories suggests that what we require is an *impersonal* standard of truth. It will be impersonal not in the sense of being unrelated to human needs, but in the sense that it is binding upon all men if they are to recognize both the needs and the means of satisfaction. Whim, edict, and feeling are alike unable to touch the standard of truth; on the contrary, the standard touches them, and is equally valid for skeptics, authorities, and men of ripe emotion, whether they accept it or not.

I think we too seldom observe the social consequences of asserting or denying such a standard, and I should like to point out what some of them are. For instance, if one denied that there is such a standard, it would be impossible then to account for the difference between honest argument and special pleading. Honest argument consists in a willingness to let one's assertions be examined according to a rule universally applicable. Such a willingness shows that one seeks no personal advantage, no exemption from a rule which others are required to obey. If there is no such rule, honesty will have to rest upon mere profession, and one need not be a cynic in order to find this unconvincing.

The most spectacular result, however, is that the denial of an impersonal standard leads to political submission and servility. For what are the alternatives? Submission follows from authoritarianism with a necessity too obvious to mention. It tends to follow from mysticism, because the cult of feeling puts one at the mercy of the reigning propaganda. As for relativism, since on this doctrine any opinion is as good as any other, one might as well choose those opinions which are safe precisely because they are submissive.

In the theory of knowledge, therefore, men have this choice: they can accept either an impersonal rule or some highly personal rulers. They can behave like scientists or they can behave like serfs. Indeed, the theory of knowledge has such profound social consequences that by asking a man's notion of it you can judge what his acts will be. You can recognize the truant, the waverer, the opportunist, the renegade, the honest man.

One can fancy Iscariot musing: "Here is my friend, who says there are certain doctrines which will make men free. And here are the priests, who say that these doctrines are dangerous. And here are the Romans, who say that the priests must be upheld. And here am I, not knowing what things are true or whether anything is true . . . And here are the thirty pieces of silver."

So the skeptic sinks into the waverer, the waverer into the opportunist, the opportunist into the renegade. It is not a question of natural depravity. It is simply the effect of social pressure upon no standards.

These considerations are the more important because a lack of standards can, with skillful embroidery, be made to pass as freedom of mind. Unless I misinterpret recent events, a theory has developed (or has been developed) to the effect that the more you waver, the freer you are. An astounding line of renegades has passed across the committee rooms, the courts, and the newspaper pages, on the road to infamy. It is as if our rulers, like Falstaff, had "unloaded all the gibbets and pressed the dead bodies. No eye hath seen such scarecrows."

The burden of their message is that, whereas in their greenish youth they were slaves to doctrine, they now in their yellow age think as they please. At least half of this is true, for they are plainly pleased by their present thoughts. Yet emancipation from doctrine seems hardly accomplished by throwing the doctrine out and replacing it merely with profitable opinions. If the deserted doctrine was true, there was some loss in deserting it; and if it was not true, there will be no *intellectual* gain unless true notions take its place. But a renegade is a man in panic, not a scientist seeking truth.

The claims to freedom of mind, however, are not limited to renegades. They rise across the whole peninsula which snugly separates the center from the right. And with them rises the denunciatory cry that men who hold views to the left of Mr. Truman are lacking in freedom of mind.

The area to the left of Mr. Truman is large. There are a good many opinions, and many good opinions, in it. But none of these is such that the mere expression of it would be likely to increase one's wages or social prestige or favor with the authorities. Consequently these opinions cannot be held on opportunist grounds. There seems, indeed, to be small reason for holding them except that one believes them to be true. If this is slavery, let it be so named. But a freedom of mind which consists in holding safe opinions seems somehow to be less than freedom and less than mind.

To deny standards is to affirm submission: events have written this maxim large enough for all to see. With it goes the corollary: to deny submission is to affirm standards. If we are noble enough to resist unseemly pressures, and if we want our resistance to show something more than personal caprice, then we must suppose that there is a standard of truth which is independent of classes and governments and capable of testing all that rulers say. Without this it is quite impossible on rational grounds to carry out the smallest reform, for every social change which trenches upon the power of rulers involves an assertion that the rulers were mistaken about either the facts or the values of the existing world. Otherwise the change is mere power politics, unenlightened by any play of reason.

Your true reformer, then, or revolutionary, will agree with Milton: "I preferred Queen Truth to King Charles." Rulers can be put to various tests by their exasperated subjects, and even the authoritarian principle can be used against them: "Rebellion to tyrants is obedience to God." But the test which is fairest, being valid for all, and strongest, being subject to no whim, is the test by an impersonal standard of truth. Only this can make a science out of politics, which otherwise must be an arena of dark forces, darkly understood. Only this can assure Mrs. Nixon that her helpers are really helpers and not teasing, pleasing, wily comforters of Job.

Being True and Being Known

From time to time I have used the noun "truth" as a convenient substitute for the more accurate but clumsy phrase "true statements." There was a danger in this of suggesting that truth is a "thing," whereas in fact it is a quality and one which belongs only to statements or sentences or propositions (we shall take these words synonymously).

Let us say, then, that our present task is to define the adjective "true." In ordinary usage it is possible to speak of a true (i.e. loyal) friend and of "the true, the blushful Hippocrene" (i.e. the genuine stuff). But neither of these senses concerns us here. We are discussing human knowledge, and knowledge is made up entirely of statements which have the remarkable quality of being true.

Now, as to the relation between this quality and the various statements men make, there are three things I shall chiefly want to say. And in order to give the discussion as much clarity as I can, I propose to write down the following sentences, which can then serve as examples:

(*a*) The Declaration of Independence was signed in the year 1776 A.D.

(*b*) George Washington was the eldest son of Louis XIV.

(*c*) Herbert Hoover is President of the United States.

Thus equipped, we shall proceed to the three main comments.

(1) A survey of the examples shows (what we already knew) that some statements are true and others aren't. It is true, for example, that the Declaration of Independence was signed in the year 1776 A.D., and it is false that George Washington was the eldest son of Louis XIV. Evidently, "true" denotes a quality which statements acquire only under certain conditions. If we can determine what these conditions are, we shall know what it means to call a statement true.

(2) Reflection upon the examples (we shall use all three this time) will also show that a true statement never ceases to be true and a false statement never ceases to be false. Thus it is "eternally" true that the Declaration of Independence was signed in the year 1776 A.D., and it is "eternally" false that George Washington was the eldest son of Louis XIV. The adverb, which I have put in quotation marks so as make it less portentous, means simply that if the Declaration really was signed in the designated year, nothing can happen that would alter the fact; and that if George Washington really wasn't the eldest son of Louis XIV, nothing can happen which would make him so.

Conceivably, books might be written to show that the Declaration was signed in some other year and that George Washington *was* the eldest son of Louis XIV; but such arguments, assuming them to be successful, would only prove that the first statement never had been true and that the second never had been false. There is no argument which could prove these statements true at one time and false at another.

This may seem surprising, because we do sometimes get the impression of a statement's changing with respect to truth. When this happens, it is because the statement contains a variable, and will be true or false depending on what is substituted for the variable. The third example illustrates this. "Herbert Hoover is President of the United States" is false if it refers, say, to the year 1950 A.D., but true if it refers to 1930 A.D. It thus resembles an algebraic proposition like $2x = 6$, which is true if $x = 3$ but is otherwise false.

The variable in "Herbert Hoover is President of the United States" is hidden because we usually don't burden sentences with all the terms which would be required to express the whole meaning. That is to say, we usually leave some of the meaning implied, and particularly we don't follow the cumbersome practice of indicating the time reference down to the day, hour, and minute. In "Herbert Hoover is President of the United States" the present tense of "is" would mean, to most people, "at the present time." That, however, is a psychological, not a logical, implication of "is." Consequently we shall fare better the more we accustom ourselves to make the whole meaning explicit. Learn to remove the variables from sentences, and you will find that the sentences you then get are, if true at all, always true, or, if false, are always false.

Hence, nothing can destroy the absolute truth of a true proposition or the absolute falsity of a false one. If you are a person who fears absolutes — a much admired but curious state of mind — you may as well learn the disturbing fact that there are a great many absolutes to be feared. This is why truth is objective and impersonal; this is why governments cannot alter it nor propagandists invent it as they please. Men may by their own efforts determine how many true statements they will know to be true: they may cultivate science or cherish myth. But they cannot possibly make a true statement false or a false statement true. In this respect they are powerless, and it is the greatest of blessings that they are so. Otherwise, *Mein Kampf* would be a work of superlative knowledge, and the *Pisan Cantos* would be an oracle.

(3) The third observation we have to make is this: statements can be true or false without anyone's knowing which they are. For example, "The earth is a sphere" was true during long ages when it was thought to be false, and "The earth is flat" was false during long ages when it was thought to be true. There is even a sense in which you could say that a statement is true or false without anybody's formulating it at all. "Lightning is a discharge of static electricity" was surely true in ages when men hadn't the faintest idea of static electricity and so couldn't frame the proposition; and it was also true, I suppose, in ages before there were any animals who could frame propositions.

It seems quite clear, then, that the conditions under which a statement is true are different from the conditions under which a statement *is known* to be true. Very probably the difference is that human beings are not necessarily involved in the first set of conditions, but are necessarily involved in the second. We shall work out the implications of this a little later. Meanwhile I want to deal more extensively with the fact that there is a difference between being true and being known.

There have been many theories which deny this difference, but they are either not frank enough or not discerning enough to say that the consequence of this denial is the impossibility of error. For suppose we deny that there is any difference between a statement's being true and its being known to be true. Then it follows that these two conditions are identical, and that whenever a statement is true, it will be known to be true. I have already suggested that this is contrary to fact, and that throughout history a great many statements have been true without being known to be so. But the point comes home more powerfully when you perceive the consequence that no one can ever be mistaken.

Error consists in believing a statement to be true when in fact it is false.[3] But the falsity of the believed statement is equivalent to the truth of its contradictory. Therefore, whenever I entertain a false statement, there is always at least one other statement which is true without my knowing it to be so. If, however, this is said not to be the case, then the consequence is that I can never entertain a false proposition. In other words, I can never be mistaken. It seems an easy way to omniscience.

Suppose, for example, I believe that Queen Victoria died in 1860. This belief is, of course, false. Now, as a matter of simple logic, the falsity of "Queen Victoria died in 1860 A.D." is equivalent to the truth of "Queen Victoria did not die in 1860 A.D." Consequently, "Queen Victoria did not die in 1860 A.D." is true, although I

---

[3] You can say that error also consists in believing a statement to be false when in fact it is true. Believing a statement to be false, however, is equivalent to believing its contradictory to be true. By substitution of equivalents, accordingly, this second kind of error can be expressed as the first.

don't know it to be true because my notion is that she did die in that year. Unless all this is the case, I don't see how I can possibly have been mistaken.

Of course, one might say that even though *I* don't know a particular true statement to be true, someone else does, or God does, and in that sense being true and being known are still the same. Of these two possibilities we shall discard the supernatural one for the reasons set forth in Chapter 5. The other possibility might very well hold as regards any individual man and *his* knowledge. It would, however, make group ignorance inexplicable, and leave us powerless to describe the evolution of mankind from superstition into science. And I think, finally, that if being true and being known to be true are the same thing, however this identity is interpreted, we shall find it very difficult to understand how anyone can even be in doubt.

What Is Truth?

In this, as (I hope) in other things, we shall decline the role of Pilate, and stay for an answer.

If the discussion thus far has been correct, the conditions under which a statement is true are not the same as the conditions under which a statement is known to be true. The theory of knowledge, accordingly, falls into two parts: the problem of defining the criterion, and the problem of methods for applying it. That is to say, if a statement satisfies the criterion of truth, then the statement is true; and if we can successfully use some method for determining whether the statement satisfies the criterion, then we shall know that the statement is true. The first of these is now to be discussed. The second belongs to the chapter following.

A statement (or sentence or proposition) is an arrangement of symbols (usually words) according to the rules of syntax of a particular language. The words refer to something other than themselves,[4] and the syntax refers to something other than itself. Thus in the sentence "John is kissing Mary," "John" refers to some specific human male, "Mary" refers to some specific human female, "kissing" refers to a specific action, and the entire syntactical relation of subject-verb-object refers to a specific relation, which we will assume to be pleasant, existing between John and Mary in space and time.

Perhaps we should say what we mean by "refers to." Reference is an act by which the sentence evokes some kind of behavior toward the given situation. The behavior evoked will at the very least be attention, but it may be more. Thus, "John is kissing Mary" is at least a way of saying, "Oh, look! John is kissing Mary." And in an irate parent, "Gad! John is kissing Mary!" would show that more than attention is being evoked.

Thus sentences are one of many links between man and his environment. Now, suppose I ask you, under what conditions would you take the sentence "John is kissing Mary" to be true? Your reply will be — will it not? — that "John is kissing Mary" is true if John actually is kissing Mary, but that it will be false if he isn't. And by this I think you will mean that John's kissing

---

[4] Except for self-referent statements, which are the source of many paradoxes but which do not, I think, necessitate any changes in the argument here advanced.

Mary is an event occurring in space and time quite independently of any sentence about it or of any observation of it by other people — and indeed more likely to occur when it can have this independence.

The conditions under which a statement is true involve, therefore, three elements: the statement, the reference to a situation in space and time, and the situation itself. We can say, then, that when the situation exists, the statement referring to it is true; and when the situation does not exist, the statement which refers to it as existing is false. If this language seems overcareful, we can try a more colloquial form: When things are such-and-such, and a statement says that that's the way they are, then the statement is true. Or, right you are, not necessarily if you think you are, but if the world *is* what you say it is.

In the light of all this, I think we can understand why we can't make a sentence true simply by feeling about it in a certain way or by asserting it with great authority. Neither intensity of feeling nor power of prestige will alter the fact that the given situation is just what it is at the given place and time. We may alter the situation by acting upon it, but then we get a new situation at a new time, and any sentence referring to this new situation will be a new sentence, whose truth, as before, we have no means of altering.

Events are somewhat in our power; truth is not. I have said that this is a blessing on political grounds. It is so on every ground one can imagine. For our control of events depends upon our knowing what they are, that is to say, upon our passing beyond fancy and feeling and fiat to the events themselves. If it were not for the tact that we cannot alter truth, it would be quite impossible for us by conscious deliberation to change the world.

It seems clear that any other view of the question must be subjectivist. It will regard truth as dependent, not on a relation between the sentence and the objective fact, but on a relation between the sentence and the human knower. It will hold the sentence to be verified, not by what goes on in the world, but by what goes on in the mind. It will be dragged downhill by the enormous gravitational force of solipsism, to hide forever at the foot, surrounded by mist and dreams.

I had got this far in my meditations, some months ago, when I chanced to read the remarks of two of baseball's most celebrated umpires concerning their own methodology. So reading, I was struck with surprise at a thing really very obvious: that umpires cannot work without being philosophical. Behind their masks and chest protectors they are forced to contemplate the great problems of existence, and they arrive (as I might have guessed) at the usual philosophical positions.

Specifically, it seems, they disagree on the question how far their decisions depend upon the outer world and how far upon inner sensation. Bill Klem, perhaps the most august of arbiters, once declared, "I don't call them as I see them; I call them as they are." This is a triumphant proclamation of the view which I am advocating in these pages.

On the other hand, Charlie Moran, an umpire scarcely less insuperable, entertained Bishop Berkeley unawares. Said Charlie, "They may be balls and they may be strikes, but, until I call them, they ain't nothing." The first clause concedes more objectivity to events than Berkeley would have approved; but this is much softened by the subjunctive mood, and is overwhelmed by the majesty of the conclusion. *Esse est affirmari:* "Until I call them, they ain't nothing."

No doubt Charlie Moran meant to vindicate the authority of umpires, and, in a pragmatical sense, he was perfectly right. So far as the play of the game is concerned, the pitches will be what he says they are; and dissent can be quelled, as in politics, by banishment of the dissenter. Undoubtedly this saves the game from rhubarb and from blows. A practical need begets the authoritative person, whose judgments thus become valid for all. The authority, trusting his own sensations and his convictions about them, ends by thinking that they are as final for knowledge as they are for the game itself. It is pleasant to see that Charlie Moran escaped this last illusion, to which many an emperor and princeling has fallen prey.

For it seems clear enough why umpires exist, and the reasons, when they are assembled, will vindicate Bill Klem. It is assumed that a ball thrown from pitcher to catcher has a certain trajectory, and that that trajectory either will or will not pass through a column of empty space bounded horizontally by the knees and shoulders of the batter and vertically by perpendiculars erected around home plate. If the trajectory of an actual pitch lies through this area, the pitch is a strike; if not, it is a ball.

The umpire's primary task is to observe and report the trajectory of every pitch. For this purpose he stands, admirably armored, behind home plate. He stands there because that position in space is the one most likely to yield accurate observation. But if the flight of the ball is not an event occurring in a particular place at a particular time, if it is merely something in a private consciousness and thus arbitrarily determinable, then the umpire might as well sit, much more comfortably, in the left field bleachers.

These are the philosophical assumptions of baseball, and they seem hardly to be doubted. When the spectators, from their high but not impartial seats, recommend that the ump be given a new pair of glasses, they seek to improve, not his observation of his own soul, but his observation of the external world. When the batter, after a called third strike, points to a position half way between third and home, he means to say that *that* is where the pitch traveled, rather than through the empty column over which he stood guard.

The belief behind baseball, behind every other game, and indeed behind all human activities except those of extreme sophisticates is that the behavior of things does not necessarily depend upon our observation of them, and that what we say about the world is true just so far as we are able to state what actually happens. Human beings have this strain of natural and (I hope) incurable realism. If they were at all times free from the pressures of organized mythology, if they proceeded upon nothing but science and their own common sense, their rise in knowledge would be astonishing, and their control of events would set them at peace with one another and with the world.

CHAPTER VIII **THERE IS METHOD IN'T**

WE HOLD human knowledge to be a relation among men and sentences and states of affairs. When the state of affairs is what the sentence says it is, then, according to our theory, the sentence is true. And when the true sentence is believed to be true by someone, that person, we say, possesses knowledge.

Between the man and the sentence, then, there exists the relation of belief. Between the sentence and the state of affairs there exists the relation of "correspondence." The relation between these two relations is knowledge. Thus, "I know that the front door is locked" means: "I believe that the sentence 'The front door is locked' corresponds to the actual state of the front door, and moreover it does so correspond."

We have held, also, that there is a difference between being true and being known to be true. Throughout human history, we said, there have been a great many sentences which were true without being known to be true; and moreover, unless a sentence can be true without being known to be true, or false without being known to be false, error is impossible. For error is the relation of belief between a man and a sentence at just that moment when the sentence does not correspond with the actual state of affairs.

The problem of all knowledge is how to make belief in a sentence coincide with the correspondence of the sentence to fact. The problem is partly ethical, since you must choose to believe such a sentence whenever you can find it. Finding it is the other part of the problem, and this too, as we shall discover, has moral concern. At all events, we have now to ask, how can we recognize true sentences; how, that is to say, can we know when a sentence corresponds with fact? We have spent three chapters in showing the necessity of a standard, in eliminating the irrational candidates for that honor, and in deriving a standard of our own. Our question now is, how is that standard to be applied?

This is the problem of scientific method, and there are two or three preliminary remarks we ought to make concerning it.

(1) The question of applying a standard is of course intimately connected with the question of what the standard is. Nevertheless, the two questions are not the same, and, if one is resolved into the other, the result must be either a loss of the standard or a loss of the method. The denial of method is surely one of the rarest theories in philosophy and amounts to pure intuitionism — anything from hunches to supernatural illumination. The resolution of standard into method, however, is far from rare and is perhaps the greatest single curse now visited upon philosophy. It is the new, immensely sophisticated solipsism, which suggests that the universe is nothing and the method is all. Behind the walls of methodology you can live as secretly and blankly as

ever a Leibnizian monad within its isolated self. But this practice is like trying to alter a landscape by tinkering with the motor of your car.

(2) If we take method and standard as dialectical opposites, we can say that the principle just formulated emphasizes their "strife." We may now describe their "unity." When you ask how, as a matter of historical fact, men refined their methods of inquiry, the answer is that they did so by observing their own successes. The advances of scientific method in the seventeenth and eighteenth centuries obviously follow upon the achievements of Copernicus, Galileo, and Newton.

At the same time the achievements could not have occurred without some previous refinement in methods. Science seems caught in a dilemma, in which knowledge cannot increase unless methods are refined and methods cannot be refined unless knowledge increases. Happily this state has been the opposite of paralyzing. I suppose that the earlier scientists, having caught one or two improvements in method, were able, like Kepler, to snatch great truths out of quite fanciful conjectures. Then later thinkers, contemplating the process, were able to elaborate the methods which had been used. But, for all that, human knowledge does give a strong impression of pulling itself up by its bootstraps. Nevertheless it pulls.

(3) Methodology (that is to say, the set of procedures by which knowledge is acquired) is strongly influenced by the structure of society. For instance, it seems hardly accidental that the characteristically Greek contributions to knowledge should have been geometry and deductive logic. It is notorious that the cheapness of slave labor delayed for long ages the development of

machinery. Now, machinery requires, in order to be invented and produced, a vigorous experimentation with the physical world; and experimentation begets a kind of mathematics and a kind of logic which are inductive and suited to the description of change. Geometry and formal logic were, however, natural companions to what we may call the contemplative attitude toward the world, the attitude of leisured gentlemen whose physical wants were supplied by slaves and whose interest in the universe was largely that of spectators.

The conclusion is irresistible that the Greeks could no more have produced calculus and (despite Aristotle's shrewd comments) inductive logic than they could have produced the electric light. By the same token, modern society could no more rest content with the four syllogistic figures than with the wooden plow. It is very possible for one man not to know what he is doing, but the main transactions of society cannot take place unobserved. The production and distribution of goods, which are the most essential of activities, necessarily lead to a detailed study of procedures, and this study is what constitutes the theory of knowledge in any age. Accordingly, there *are* such things as slave science, feudal science, bourgeois science, and socialist science, in the sense in which these phrases designate the state of human knowledge in particular historical epochs. The phrases may be applied abusively, but they have a literal meaning for all that.

Human knowledge is a *social* fact, and any theory of method needs to take account of it. Society, indeed, affects knowledge in two ways: it limits or enlarges the chance of new discoveries, and it conditions the observer as to what he will believe. The scholastics who (it is said) rejected the plain sensory evidence of Galileo's experiment were no doubt victims of an erroneous theory, but they were also victims of a moribund social structure. Error in itself is just error; compounded with social lag, it is prejudice.

We shall recur to this fact when we discuss "objectivity." Meanwhile, let us proceed to the problem of method as such.

The Sensible, the Thinkable, and the Practical

If we are to determine whether a sentence expresses the actually existing state of affairs, we have to have some kind of direct contact with that state of affairs itself. We must, so to say, be able to hold the sentence in one hand and the world in the other and examine them both. In order, for example, to determine whether the sentence, "There are two books on my desk," is true, we must establish a relation between ourselves and the books on the desk.

Our first appeal, of course, is to sense experience. We look at the desk, and if the books are there in the asserted number we accept the sentence as true. In this respect we are all children of our father Locke. Indeed, we have always been so. Locke's epigram about Aristotle might be rephrased to apply to himself: "God has not been so sparing to men to make them barely two-legged creatures, and left it to Aristotle to make them rational."[1]

[1] *Essay*, Bk. IV, Chap. XVII, Sec. 4.

For God did not leave it to Locke to make us empiricists. Rather, I am sure that our ancestral ape, indicating by grunt the existence of a tree, verified the grunt by looking at the tree, by touching it, or, more triumphantly, by climbing it.

The assumption behind these empirical predilections is that our senses are avenues directly connecting us with the world. It may be that in skeptical moments we wonder whether the avenues have become road blocks and we may even seriously entertain the sophisticated difficulty of distinguishing between visions seen and visions dreamt. But we have a radical conviction that whatever we can know to exist must manifest itself to our senses, at least in its effects, sometime and somewhere. We are not perhaps so bold as to claim inerrancy for all our perceptions, but we do believe that whatever we know has to have some foundation in our sense experience.

Undoubtedly we came to this more cautious view through a lifetime of hard knocks. Sense experience, it now seems, is a necessary but not a sufficient condition for knowledge; that is to say, with it we still don't necessarily know, but without it we are quite certain not to know anything. I might look at my desk and cry out, "See, there *are* two books there!" and still be mistaken. There remains some risk of error, though the risk is small. What I cannot do is get any verification of that sentence by purely nonempirical means. I can't deduce it from any premises unless at least one of the premises is empirical. I can't get it by mystical insight or by authority. Somewhere and sometime I have to consult my senses about it.

Empiricism has many difficulties, but none of them is such as to do away with sense experience as one of the conditions of knowledge. The difficulties, however, do suffice to show that sense experience is not sufficient for the purpose. Let us examine some of them.

(1) There is the familiar fact of illusions, mirages, hallucinations. These occur empirically, but are recognized in time to be distortions of reality — "mutilated and confused ideas," as Spinoza liked to say. Now, in a certain sense, the mirage of a water hole is precisely what the desert wanderer "ought" to see under the given conditions of light and eyes. If this is what we mean, then it will be the case that "the senses never lie." But obviously the kind of vision which presents a water hole where none exists has something erroneous about it. You might say that the trouble lies in the interpretation which the wanderer puts upon his sense data. This, however, is a clear admission that more goes into the making of knowledge than the sensation itself.

How are these sensory illusions corrected? Well, the least you can say is that they are corrected by later sensations. The wanderer comes up to the site of the supposed water hole and now *sees* no water hole there. Why does the later sensation refute the earlier, rather than the earlier the later? Because of some rule or other, which says that accuracy increases with nearness or with evidence from other senses or with the predominance of one set of sensations. But the rule is not itself a sensation. It (or something like it) is nevertheless quite obviously necessary for knowledge. Consequently more than sense experience is required if we are to know.

(2) Perhaps the greatest difficulty, however, is the one which Hume pointed out. He took it to be a defect in knowledge rather than in his own theory, but the difficulty itself is real enough, whichever it applies to. It is, namely, the fact that our senses tell us surprisingly little about relations. Relations in space they do undoubtedly record — and relations in time, unless we are to suppose that sensations are a series of disconnected pictures like the stills on a movie reel. But of relations other than the spatial and the temporal the senses give only hints. In particular they do not show whether, as between the pulling of the trigger and the firing of the bullet, there is anything more than simple succession in time.

A philosophy basing itself on sensation alone, like positivism, consequently has extreme trouble in accounting for human knowledge. The difficulty has been admirably stated by Whitehead:

Suppose that a hundred thousand years ago our ancestors had been wise positivists. They sought for no reasons. What they observed was sheer matter of fact. It was the development of no necessity. They would have searched for no reasons underlying facts immediately observed. Civilization would never have developed. Our varied powers of detailed observation of the world would have remained dormant. For the peculiarity of a reason is that the intellectual development of its consequences suggests consequences beyond the topics already observed. The extension of observation waits upon some dim apprehension of reasonable connection. For example, the observation of insects on flowers dimly suggests some congruity between the natures of insects and of flowers, and thus leads to a wealth of observation from which whole branches of science have developed. But a consistent positivist should be content with the observed facts — namely, insects visiting flowers. It is a fact of charming simplicity.[2]

It is indeed — and a little too simple for science. For if there were no more to knowledge than sensation itself, we should live upon the surface of events like flies on water, and should catch only the momentary shadow, the brief flicker of light. We could not construct a single generalization, and though we might discern movement, we could never catch the intimacies of its growth. Life, a dome of many-colored glass, would not carry even a poetic imagination to the infinity beyond.

The truth is that we know with the whole body and not just with the senses. It takes the brain and nervous system for a sensation even to complete itself. But the brain and nervous system seem also to be the source of that "intellectual development of consequences" which Whitehead shows to be imperative for science. I do not know whether the cerebral cortex is in fact the great canonical organ of all our logical activity, but I am quite sure that this activity is what enables us to construct universals and use them in dealing with the novelties which pour in upon us. Memory is no doubt efficacious, but it would not be so unless we could recognize identities between the old recollection and the new state and thus select from the welter of past things those that are relevant to the present.

Take down any book of science you wish, and skim its pages. Most of the sentences and all of the important sentences are, you will find, universals. That is to say,

[2] *Nature and Life*, The University of Chicago Press, 1934, p. 24.

the sentences don't record just one event which happened at a certain time and is now, with all its individuality, vanished. They record, rather, the behavior of groups of events, and the more ambitious of the sentences refer impartially and simultaneously to past, present, and future. When you state the orbit of the planet Mars, for example, you aren't referring to just one circuit which that planet made at some one particular time; you are stating the orbit as a path of repeated travel, which indeed you expect to continue for an indefinite time.

Now, sensations (as empiricists describe them, at any rate) are singulars, and if knowledge got no farther than them, not one book of science could ever be written. There is, of course, much dispute — and quite justifiable dispute — as to how we are able to get universals out of these singulars. But that we do get them is undoubtedly the conviction of every scientist, who must therefore agree (not, I presume, very unwillingly) that in human knowledge there is an important element of sheer logical construction. It is this, and not our senses, which enables us to know that the moon has an opposite side that we never see.

Indeed, the proud empiricism of nineteenth-century science is now much chastened. The once admired glory of "sticking to the facts" has yielded to the greater glory of formulating a principle from which all facts can be deduced. It is well known that any science can be expressed as a formal system, with definitions, axioms, postulates, and theorems on the Euclidean model. This was not a victory of deduction over induction, of reason over sense, but a long-needed and long-awaited equality in marriage. The sensible and the thinkable are at last one.

But this marriage must be fertile or it will not last. What it bears are consequences. The combined constructions of sense and intellect will remain merely a way of looking at the world unless they become also a way of controlling it. This was the valid insight in pragmatism, and it would have made of pragmatism the very consummation of philosophy if the pragmatists themselves had not entered upon patricidal strife. When, however, they showed a contempt for theory and an adulation of practice, their doctrines trailed off into mere suggestions for momentary adjustment, that is to say, into opportunism.

It is ironical, but doubtless just, that in this manner pragmatism lost its chance to be practical. The Greeks had shown that you can get to know the world by thinking about it. The early moderns had shown that you can get to know the world by sensing it. There remained the chance to crown all this by showing that you can get to know the world by acting in it, where the phrase "acting in it" means the carrying out of a program based upon an entire relevant theory. The success of the program in attaining its goal is surely some proof that the theory was correct. But once pragmatism decided that theory is trifling and practice is supreme, the delicate relation of the sensible, the thinkable, and the practical in human knowledge was broken into bits. A useful object lesson, for it teaches us once more what we should never tire of learning, that opportunism is the best way to miss opportunities.

If what we have thus far said is correct, it will appear

that scientific method is an organized, systematic procedure involving sensation, logic, and practice. It seems plain that we must have all three, and that theories of knowledge have very often suffered from fastening exclusively upon one of the parts. Mere empiricists can't account for generalizations. Mere rationalists can't tell whether their deductions correspond with fact. Mere pragmatists can't know what it is their practice confirms. But if we think of verification as the testing of sentences according to observable sense data, logical coherence, and effective practice, we shall find that we have gathered into our method all the possible elements, and that the elements strengthen one another. I want to consider now, in a little more detail, what expectations such a method can yield.

Plurality and Risk

Scientific method, as a system of interrelated parts, is grounded in a certain characteristic of sentences which we have not yet described. In the previous chapter I treated sentences as individual entities, for the purpose of making clear what we mean when we call a sentence true. It is often convenient to treat sentences in this atomic manner, and indeed that is pretty much the way they are treated in the traditional logic.

But, as a matter of fact, sentences are not atomic, if by that adjective we mean that each stands absolutely alone, unrelated to any other sentence. Whenever we speak, we actually speak in *groups* of sentences, little systems of them, even though what we utter appears to be one sentence alone.

Suppose, for example, somebody says, "There is not a cloud in the sky this morning." That appears to be just one sentence. But consider, now, all the other sentences which this sentence assumes. It assumes certain sentences defining the nouns "cloud," "sky," and "morning"; certain other sentences defining "there is," "not," "in," and so forth. In short, anything you say, however singular it may appear, will assume a number of other statements having to do with the syntax and vocabulary of the language you happen to be using.

Now, definition is the act of establishing the reference of terms, and reference, as we said some time ago, is the directing of behavior toward the environment we live in. Moreover, since language is a social possession (unless you have invented one entirely your own), definition is a social act. It means that the users of a language, who may of course number millions, are trying to direct their behavior toward environment in as nearly as possible the same way. And indeed, unless this unanimity can be at least approximated, there won't be any social action or even any living together.

What is true of vocabulary is also true of syntax. Syntax is what enables a language to express the relations existing throughout the world. It represents an astonishingly high level of generalization, because the vast number and variety of particular relationships can all get themselves expressed by prepositional phrases, subject-predicate structures, and the like. Syntax is thus the anatomizing which mankind has performed upon the skeleton of the universe — or, we may rather say, is still performing, since the knowledge gained by previous analysis refines the analysis itself. It is the reflection of cosmic structure in syntax which rescues us from the isolation of immediate circumstance and enables us to act upon a deliberate plan.

In addition, however, to those other sentences which any uttered sentence must grammatically or logically assume, there is yet another group bound to it by simple association. Here we shall find what may be called "emotional overtones," suggestions of praise or blame, of joy or sorrow, of value in general, or of mere random recollection. Although this is supposed to be peculiarly the stuff of literary expression, it is difficult, and perhaps impossible, to rid any sentence of it completely. I should imagine that most people, confronted with the sentence, "There is not a cloud in the sky this morning," would feel the exhilarated ease that goes with that sort of weather. The connection is of course not logical or grammatical, but it can hardly be escaped.

We must say, then, that whenever we talk, we talk in volumes. Loquacity is simply the resolute, and perhaps neurotic, uttering of all those sentences which the saner or the more controlled are willing to leave implied. Now, the volumes we talk in get some of their content from sensation, some of it from that kind of thinking which shows us the joints of the world, and some of it from the practical activity in which we are always and inevitably engaged. It follows that we cannot say anything without assuming that unity of the sensible, the thinkable, and the practical which we have declared to be the essence of scientific method.

When we were meditating, a moment ago, upon the sensible, the thinkable, and the practical, we observed that there are great defects in each if taken separately.

No *one* of them, we said, would suffice for knowledge. We thought, however, that, taken collectively, they would suffice.

On this view we must now place some qualifications — such, specifically, as will show what we are to mean by the word "sufficient." We have held a sentence to be true if the facts are what the sentence says they are. We have held that every sentence, provided it is free from variables, is either true or false and always remains whichever of those two it happens to be. Thus far we speak and, as I think, must speak in absolutes, for the ground of all knowledge is this kind of articulation between sentences and the world.

But when we come to describe our *recognition* of true sentences or false sentences, then our human limitations begin to appear. Restricted by immediacy to a small area of time and space, we have nevertheless to talk about far greater spaces and times. Confronted by a welter of qualities, we have to feel out conjecturally the system which gives them form. Moving into the future and its novelties, we have to test by action all that we, by brain and senses, took to be the case.

Thus, whatever our certitude that a given sentence must be true or false, we shall be left with a modicum of doubt which of the two the sentence is. You may say, for example, that the length of a table at a given time either is or is not precisely six feet. The length of the table, however, varies with temperature and pressure, and the same is true of the length of your measuring rod. Accordingly, as a practical matter, you take a series of measurements, and these will show you the limits, doubtless quite narrow, within which the length varies.

Now, of course, you can't spend your whole life measuring that one table. You are obliged, therefore, to adopt what is called the small-sample technique and decide the length as best you can on the basis of a few measurements. In this there is some risk of error, but you have to run it. As you run it, you may be comforted by the marvelous beauty of statistical procedure, which, while it alarms you with the prospect of error, eases the soul by stating with mathematical precision what the chance of error will be.

The chance of error is the chance that a given sentence does not express the existing state of affairs. When this chance stands in relation to human purposes, it becomes a *risk,* because it indicates a possible defeat for behavior based upon accepting that sentence as true. If I order a meal in a certain restaurant, I do so in the belief that the sentence "This meal will nourish me" is true. Human experience of restaurants shows that that sentence cannot be known with absolute certainty, that consequently there is some chance of error and some risk in accepting it. But if I continue immobilized by doubt, I shall soon feel the sting of Locke's maxim: "He that will not eat till he has demonstration that it will nourish him, he that will not stir till he infallibly knows the business he goes about will succeed, will have little else to do but sit still and perish."[3]

What I have to do is to make sure that the chance of error decreases as the loss from error increases.[4] If I am stricken with a dangerous disease, it is rational of me

---

[3] *Essay,* Bk. IV, Chap. XIV, Sec. 1.
[4] This whole discussion owes much to Professor C. West Churchman's *Theory of Experimental Inference,* New York, Macmillan, 1948, Chap. XV.

to consult a physician rather than a witch doctor, because the chance of error in the physician's diagnosis and treatment, taken in comparison with my possible loss, is far smaller than the chance of error in mere incantation. In other words, the risk is smaller, and the smaller risk helps to determine what I hold to be knowledge in this case.

This, I take it, is one of the ways in which practice molds theory. It obliges us to treat our descriptions of the world as studied risks involving gain or loss for human purposes. It obliges us to conduct science as an activity and not as a kind of contemplation. Moreover, since practice is always in some manner social, the calculation of risk will turn in the end, not upon this or that need of this or that person, but upon the universal needs of mankind. In this way, as Marx suggests, "all mysteries which mislead theory to mysticism find their rational solution in human practice and in the comprehension of this practice." [5]

Now, every decision about risk is a value judgment, since we are choosing one statement rather than others on the ground that it is least likely to be erroneous and therefore least likely to defeat our aim. It follows, then, that practice introduces into knowledge an ethical element which sensation and logic might seem to lack and which the purely empiricist or purely rationalist theories of knowledge do in fact ignore.

Such ignoring is ignorance, and though it may not be willful, it is nevertheless complete. For what would be the point of evolving an intricate and careful methodology, with all the elements of sensation and logic

[5] *Theses on Feuerbach,* No. VIII.

and practice in their proper places, if one chose to reject all statements thus derived? Such a choice would doubtless be perverse, but it could be made — indeed it is made daily by very many people. One accepts the probably true statement because one prefers it, and one prefers it because one deems it more valuable. The truth of a statement (its correspondence with fact) does not necessarily involve ethics, but our belief in the true statement, our acceptance of it, always does so. You cannot say that oceans exist because they are good for steamship companies, but you can say that the statement "There are oceans" is, in view of the evidence, worthy to be believed.

Thus ends the divorce between fact and value, which, of all present illusions, has corrupted thought the most.[6] The scientists who think they can know without valuing, like the artists who think they can value without knowing, are profoundly deceived. Doubtless this fancy heals the conscience when bent upon a deadly enterprise. But the elimination of ethics from science means the elimination of practice from method. Theory, thus severed from its use, becomes a sterile contemplation in which things are seen not quite as they exist and the existing things are not quite seen. Such contemplation might very possibly be pleasant and even serve the random needs of fantasy. But who would call it science?

[6] We shall recur to this point toward the end of Chapter IX.

SUCH A HARD TIME

CHAPTER IX THE CANDLE
AND THE SUN

A YOUNG COLLEAGUE OF MINE, engaged upon the instruction of a class in ethics, lectured one morning about the nature of conscience. He threw the subject open for discussion, and there ensued the customary pause during which students ponder what they had better say. Then the fairest and most disputatious of his coeds remarked that, as for conscience, she didn't think she had any. "What," said the voice of a male student, "are you doing next Saturday night?"

It was some time before the laughter and the blush subsided; but when philosophizing became possible once more my friend realized that the two students had between them demonstrated more about ethics than any lecture could possibly do. The girl had only intended disbelief in a certain spiritual faculty. The boy had suggested that a person limited by no principle will be limited merely by his own powers or society's. Between the one boundary set by ethics and the other boundaries set by nature or the police there is a realm of charming possibilities such as it takes a week end to explore.

In a way, the history of ethics is a contest of these boundaries. The police, or at any rate administrators who employ them, will usually maintain that legal limits and moral limits are identical — a view which enables them to say that what they punish as illegal is justly punished as bad. Sociologists incline to a similar view, but they mean by it something quite different. They mean that morality is so completely a social code that you can't tell what is bad until you know what the police will punish.

What may be called the theory of natural morality has also its extremes. You may hold with the Stoics that the universe is justly constituted and that its happenings, whether pleasant or painful, are to be accepted with serenity. Or you can say with the younger sophists that custom is artificial and nature "natural," and that in any contest between the two ethics one should side with nature against the tyranny of codes.

I suppose there can be a third view, possible to supremely confident societies, in which the limits of law, nature, and morality will coincide. An Elysian concord this would be, where nothing thwarts the even flow of wishes, and every end is gained at no one's loss. Perhaps we shall find that in a certain reading of this concept lies our ideal.

As one contemplates the present state of ethical theory, one feels some pathos and some exasperation. One expects disagreement in the definition of an ideal: such things escape precision while at the same time inviting it. But it is quite another matter that the definitions should be so different and even opposed, that schools of moralists should war upon other schools, that on the most precious concerns of life a kind of anarchy should flourish in human opinion.

Our century, moreover, has seen two prolongations of the anarchist trend. One is a theory which holds that the word "good" is indefinable, and from this theory it follows that men won't know whether they are using the word in a common sense. The other is a theory that moral statements express merely personal commands, wishes, or tastes.

Now, we said in our first chapter that the reality of change, its control by knowledge, and our awareness of possessing this control would all be futile unless we could display some goal toward which the entire process ought to be directed, and thereby establish some rule for the governing of choices. This was to have been the crown of all our work for Mrs. Nixon, the solace and the glory of thought made meaningful for man. But entering now the final sanctuary, we find its beauties veiled and its clear silence shattered with contrary vociferation. Can it be that the builders mistakenly built Babel? Or, as their critics say, did they build nothing at all?

It may appear that, whatever the doubts of moralists, there can't be much question what Mrs. Nixon needs and ought to have. If you could confer economic security upon her, you would do so and rejoice in your act without exhausting the details of ethical theory. An incompleteness in the theory does not require hesitation in the act. But if you will view yourself as more than a casual benefactor, as a social planner (which in some degree you are), then the inevitable bond between action and theory becomes manifest. You now find yourself considering whether men do really share a common goal and what, if anything, that goal is.

Moreover, you will find that, as the theory solidifies into a program, there will be hostile voices attacking either its practicability or its morality. These are the classic methods of opposition: the one says it can't be done, the other that it shouldn't be. There is some comfort in observing that the two attacks, launched simultaneously, will negate each other; for if the thing can't be done, it doesn't much matter that it shouldn't, and if it shouldn't be done, we are much eased to know that it can't.

Attacks upon practicability, when they reach a philosophical level, take the forms our previous chapters have described. But attacks upon the morality of a program produce their effect by arousing compunction and diffidence in the supporters. Programs are, indeed, very vulnerable to this sort of criticism, because either their means or their ends can be questioned.

There is, moreover, the fact that programmatic criticism is not necessarily malicious. It is easy to be mistaken about the value of an end or the validity of a means. In that event, the program benefits from a discussion of its ethics. From such a discussion people learn the relation between the path and the goal, the distance (so to say) between their present place and the place desired. When people do this in harmony, they develop a common body of ethical doctrine in which the principles and the terms are very generally understood.

I suspect that the present anarchy of moralists is due in part to an unfortunate division of labor. Action is common and popular; theory is effete and academic. This division tends to make Macbeths of the agents and Hamlets of the theorists; and these Hamlets, lamenting that they were born to set anything right, ceaselessly prosecute their skeptical errands. Perhaps the ghost lied or was illusory. There must be experimental proof: "the play's the thing." "Now might I do it pat!" — but, no, the King is too near heaven. Thus the theorist sinks toward madness and finds in madness a useful plausibility. Unless events throw him at last, and not too late, upon the *act,* his doubt of attainable value will have become disbelief in any value at all.

This luckless state, which yields no reason for being optimistic, no reason for being pessimistic, no reason for being neutral, and (in a moral sense) no reason for being, is the prostration of rational choice. Any culture it exists in is sick; any culture it dominates is dead. Westerners are a little too comfortable in their geography and need to realize that the deterioration of their culture has now reached a point where they must decide, not alone the ethical rule for making choices, but whether such a rule is possible at all.

"Good" Means — What?

Throughout this book our thesis has been that philosophy is the theory of human deliverance. We have supposed that deliverance is in fact desired by very many people. We shall now proceed to hold that deliverance is worthy to be desired and worthy to be attained, not just by those who may happen to want it but by everybody. And we shall hold as a sort of corollary that philosophy is justified precisely because it makes possible the attainment of what is thus worthy to be desired and attained.

So speaking, we have followed a familiar pattern of ethical discourse. The value of philosophy, we say, lies in its efficacy as means, and that value derives from another value intrinsic in the end. By this mode of thinking, which is certainly very usual, the only *special* merit in a means is its efficiency; whatever other merit it has comes from the end toward which it is exerted. We thus get a view of human life as spread out through a series of choices, each choice being made on behalf of an end supposed to be within reach, and that end being in turn chosen on behalf of another and remoter end. Human life is thus a series of ends dissolving into means until we come, in speculation at least, to an End worthy of capitals, an End we call (or hope we may call) supreme. This is, you might say, a mountaineer's view of ethics, by which a man climbs purposefully, scorning the epicurean dance and the pragmatic dodge, until "he stands on the heights of his life with a glimpse of a height that is higher."

But the mountaineer's is not the only view. It is possible to think that intrinsic value is not hidden in some distant consummation but is discoverable from moment to moment in immediate experience. The hedonists, for example, to whom pleasure is valuable in itself, enjoy those values as the values come, broken no doubt by periods of satiety or nausea, but present when they are present and seizable as they appear.

Or, if you are the sterner sort of moralist and believe that value lies in obedience to some law, you can always realize such value without the slightest delay. In fact, you can do this much more readily than the hedonists, for you won't be interrupted by satiety. If anyone can believe this ethics and discipline himself to the constant practice of it, he can come near to gaining intrinsic value one hundred percent of the time.

Whether, in ethical theory, one belongs to the purposive, the hedonist, or the law-abiding school, one comes at any rate to the conviction that there is some single standard of value, authoritative in itself and bestowing a derivative value upon such other things as have it. Out of the great contest of human wishes there thus emerges the final question: What is the standard (if indeed there is one) and how shall the adjective "good" be defined? And it seems that if there is no standard, life will be left to those feebler meanings which private whims and social usage give it, that the adjective "good" will sink to the humbler levels of taste and etiquette.

For nearly a century, since Martineau set the fashion, discussions of ethical theory have proceeded by reviewing the accumulated doctrines and subjecting them to analysis. After this manner, I once sat down, at the end of some years' teaching, to make a list of the main ethical theories — a list such that the theories themselves should overlap as little as might be and should, in their sum, exhaust the possibilities. I arranged the theories as definitions of the adjective "good," in the way that word is used when we say, "Health is good" or "Socrates was a good man."

The list I produced was the following:

"Good," as adjective, means:
1. "Pleasant to me"
2. "Pleasant to most people"
3. "Approved by me"

4. "Approved by society"
5. "Conforming to the moral law"
6. "Conforming to the divine law"
7. "Conducive to the attainment of an ideal"

And to these seven, the main developments in ethics in our century obliged me to add the following:

8. "Good" means something, but we can't say what.
9. "Good" doesn't mean anything.

We shall now discuss these positions critically but briefly.

(1) This is the theory called (by philosophers) "egocentric hedonism." It appears to be, of all theories, the most immediate, concrete, undogmatic, unaspiring, and improvisational. When asked what is intrinsically good, it answers, pleasure. Whose pleasure? Mine; it's the only pleasure I have. Is there anything better than pleasure? Nothing except more pleasure. Accordingly, all I need do in evaluating an action is to compute how pleasant it will be to me.

The theory is not socially minded, but it is not necessarily antisocial. In Aristippus and Epicurus it was a live-and-let-live, cultivate-your-garden idea (the latter phrase is, in fact, Epicurus'). But it can have antisocial forms, as in Machiavelli and Hobbes, and in these forms it has provided novelists with much of their material. In this, I imagine, has lain its entire social usefulness.

The doctrine also has the charm of laxity and appears to offer holidays among the more genial vices. Sad to say, as a matter of historical record, the theory has been rather tame. For every swollen voluptuary it has produced a hundred Horaces, seizing the day with cautious pleasures and with a dire sense that *fugerit invida aetas*.

Old Epicurus himself, to whom is erroneously ascribed the praising of wine, women, and song, was far too wise for such sophomoric antics. He valued friendship as the only reliable union of pleasure and security, from which are exiled all the giddier delights.

Defense of the doctrine usually takes the form of a militant query, "Well, *you* like pleasure, don't you?" I have sometimes wondered what hedonists would do if I were to say, No. But, even assuming I do like it (and I do), the fact that I like it fails to establish the principle that it is good for me to like it. Still less does it establish any principle to the effect that *my* pleasure should be the object of universal solicitude.

(2) The second theory is hedonism with a social conscience, and is called "utilitarianism." It holds the relation between individuals and society to be the same as that of atoms to molecules. The moral problem always is to discover what will make the molecule happiest, and this is done by computing the pleasure or pain for each atom in the molecule and totaling the positive and negative quantities. The act giving the largest sum of pleasure, thus computed, is the act you should choose.

This is the Benthamite calculus, to which we have already referred.[1] It is not an ambitious theory. It has the liberal's lazy admiration of majorities, and one feels that its maxim about "the greatest amount of pleasure for the greatest number of people" would be amply satisfied at fifty-one percent.

If egocentric hedonism suggests the behavior of a dinner guest, utilitarianism suggests the behavior of a hostess. She has to choose the menu, and her rule of

[1] Chapter III.

choice is to please all the guests possibly and most of them certainly. For some such reason, certain articles of food, reputed to be generally and vigorously pleasant, have become staples upon these occasions. As the guests proceed from the filet to the parfait, they are probably not aware of verifying the hedonistic calculus in its commonest (and perhaps its only practicable) form. Bentham has been a skeleton at many feasts besides the centenary banquet of the Bentham Society.

The calculus has never got itself expressed in numbers, and the use of it is in fact not numerical at all. The hostess does not cover her guests with apparatus to record hedonic fluctuations. She has already anticipated and "calculated" their pleasure, partly in imagination and partly by acquaintance with their tastes. We all work, from day to day, upon certain rules of thumb concerning what either friends or strangers are likely to enjoy. Yet, though we speak of pleasure quantitatively, we don't express the quantity in numbers. Herbert Spencer once attempted something of the sort when he weighed the advantages of living in New Zealand against those of living in England. His score was: New Zealand 301; England 110.[2] But he never went to live in New Zealand.

The contentment with majority satisfactions is a bit of ethical naïveté. Why not, rather, insist upon action for the benefit of *all?* A supreme principle of morality is strangely conceived if it snuggles so close to ordinary life as to demand nothing but compromises. Doubtless commercial society, which produced utilitarianism, knew very well that it could not please everybody and

[2] *Autobiography,* New York, Appleton, 1904, Vol. I, p. 429.

considered that pleasing most would be a sufficiently spectacular feat. But we are in search of a standard valid for all historical epochs and therefore not confined to any one of them.

(3) It is possible for an ethical theory to be egocentric without in any way justifying self-interest. Theories which assert the supremacy of the individual conscience are of this sort. In their more primitive form they treat the conscience as a sort of spiritual organ, a "faculty of the soul" which delivers moral judgments.

This concept has been much sophisticated. Shaftesbury (the third Earl) turned conscience into a "moral sense," which behaved not like a justice but like a connoisseur. Bishop Butler made of it a more intellectual exercise which he called "reflection." And I think the concept survives, much thinned by analysis, in Professor George Moore's notion that we recognize "good" as we recognize "yellow" — by direct experience of it.

These varied versions have the common belief that moral value is discovered by private intuition. "Private" means that the individual agent does in the end decide for himself; "intuition" means that the value is apprehended directly and requires no proof. The most vivid reading of this theory is to be found in the Quaker doctrine of the inner light. Men in general, and the members of that sect in particular, possess each an internal luminary which glows or darkens according to the state of spiritual atmosphere. By such a view all choices are easily determined, and only the construction of programs can give serious cause for thought.

Now, it appears (to me, at least) that the lamps do

not flash at the same time or in the same rhythm or with the same color. There are some lamps which flash white for approval and red for disapproval. There are lamps which flash red for approval and white for disapproval. Even when the lamps agree upon the signal, they flash it at contrary moments. In the end, the police, by extinguishing irregular luminaries, bring all the flashings to a common time, and ethics becomes one more subject for traffic engineers.

It is always difficult to harmonize intuition with a wholesome respect for the law of contradiction. There must be some such harmony, I think, because our acceptance of the law of contradiction is very probably intuitive. But when the morally intuiting persons are set down in the midst of conflict, and their daily bread is got by strife rather than by co-operation, their lamps will flash according to their immediate needs. I think the lamps are very sensible to do so, but that is not what the theory intends.

Moral intuitionism leaves us, as all anarchist theories do, with a puzzle. If each man alone is final judge, how may common consent be required? And if judgments differ from person to person, how can that consent be obtained?

(4) The fourth theory is a favorite among sociologists, among whom it goes by the name of "cultural relativism." The doctrine is that each society develops its own peculiar morality, often set forth in explicit codes, by which it regulates the behavior of its members. As a matter of historical fact, this notoriously does happen. What is remarkable in the theory is its assertion that ethics is no more than this. To call an act "good" is

simply to say that a given society in a given state of development approves it.

The theory holds, furthermore, that codes will differ with different cultures or within the same culture at different times. This notion is by no means as new as is often supposed. It was held, I believe, by Herodotus at a fairly distant date, and in 1690 Locke let fall a decorous intimation that "the saints who are canonized amongst the Turks, lead lives which one cannot with modesty relate." [3]

Such facts are now familiar enough. But again the theory asserts its peculiar nature by holding that the historical diversity of views makes any universal ethics impossible. "Who are we to say?" ask these thinkers, especially apropos of some plan for the liberation of colonial peoples. "Perhaps the natives are happier with their present customs."

Now, the alleged diversity is somewhat deceptive. One need not be a sociologist to comprehend that, the world over, men spend most of their time in precisely similar activities. They eat, sleep, work, play, and copulate; and when you have counted up the hours given to these occupations, there isn't much time left in which people can be different. Moreover, the values attaching to these common occupations are all identical and are all shared.

The differences lie in the social relations surrounding food and work and love. To the ingenuous observer (and the disingenuous) such differences may seem large. What strange behavior! What extraordinary customs! — Anthropology can sound like the jottings of a tourist in search of quaintness. But Darwin, at any

[3] *Essay concerning Human Understanding,* Bk. I, Chap. III, Sec. 9.

rate, had the chance to know better. When he visited the primitive and impoverished natives of Tierra del Fuego, he learned with some horror that during times of winter famine they were accustomed to kill and eat their old women first and their dogs second. The explanation was, "Doggies kill otters, old women no." [4]

Can economic motive be plainer? Food is perhaps the greatest of essentials, and essential must be the means of getting it. Under Fuegian circumstances the dogs inevitably rated higher than the old women. One can imagine many an angry disquisition upon the innate cruelty of "savages." The Fuegians, however, were not cruel but only hungry. Either some must perish or most would starve.

Civilized nations do not habitually kill old women. One can say, therefore, that civilized societies of 1832 and the Fuegians of 1832 did differ in this respect. One can say that, if the Fuegians had followed European rules, they would all have perished; and if the Europeans had followed Fuegian rules, they would have committed atrocities. But I think that Europeans and Fuegians did not differ in respect of a need to eat. The values of eating were consequently common to them both.

The fact is that, throughout mankind, there is a relativism of means but a community of ends. Even the relativism of means, however, is disappearing as the world narrows. When a world society finally establishes itself, there will be a community of ends *and* of means, and the universal mores then prevailing will be far closer than we now are to the ideal.

The heart of cultural relativism is contentment with

[4] *Voyage of the Beagle*, entry for December 25, 1832.

things as things are. A slothful tolerance, a friendly inertia, wraps it in garments much tattered by inferior use. It seems never to have passed through social crisis nor even to have heard of one. For the clash of values is the very essence of crisis — a clash which obliges us, if we still cherish rationality, to fix on some one value as the arbiter of all.

(5) and (6) We can take these theories together because they have a common base. They represent in ethics the triumph of the concept "right" over the concept "good." They suggest that morality is concerned not with the satisfaction of desire but with the discipline of it. That is to say, what we want to do is of less consequence than what we ought to do, and accordingly our task — our "duty," if you will — is to submit our wishes to a universal rule.

The formulation of this rule is the chief labor to which theories of this sort devote themselves. Of all such efforts the Kantian is the most famous and the most successful. Kant undertook to determine what, if anything, in the universe can be called unqualifiedly good. Traits of character, such as intelligence, wit, and even benevolence, are capable of corrupted use. They are therefore not intrinsically good, however much one may admire them. Similarly with the objects of desire: it will be all right to desire them if they are in fact worthy to be desired.

The trouble is that I cannot get beyond that *if*. Suppose I try, taking as my subject a familiar sort of planning: I study hard in order to become a physician in order to earn a respectable livelihood in order to lead a cultivated life in order to be happy. Now, it turns out that my studying will have been worth while *if* being a physician is worth while, and that being a physician is worth while *if* earning a respectable livelihood is worth while — and so on to the last condition, of happiness being worth while. But here too is the impenetrable *if*. I may adduce further arguments to prove the value of happiness, but I can assert its value only *if* the arguments themselves are true.

A morality of ends seems therefore to lose itself in hypotheses. A categorical morality — a morality which can, as it were, be said straight out — must be sought elsewhere. Kant thought to find it in a special procedure for reaching decisions and in the discipline of human will to that procedure. Accordingly, in every circumstance of life, I am to act upon a maxim which has this peculiarity, that if everybody were to act upon it, no inconsistency would result.

Suppose, for example, I am a merchant needing a loan which I am aware I probably cannot repay. (Kant's illustrations are innocently bourgeois.) Quite possibly I could contrive to get the loan: it is what makes a good merchant. It is not, however, Kant thinks, what makes a good man. For, in acting thus, I have adopted a maxim to the effect that anybody may borrow without intending to repay; and, if everybody were to behave like this, no loans would be made and no borrowing would be possible. My maxim, taken as a general rule, is therefore self-contradictory in the sense that it is self-defeating. It cannot be a universal law.

Thus Kant made a great effort at assimilating ethics to logical consistency. It wears, as he left it, the look of artifice. But I think we must not fail to appreciate in

Kant the motive and the hope. He sought an ethics which, being strictly valid for all men, could not be twisted into special pleading. He wanted to demonstrate an equality of rights and duties, and a brotherhood such that everyone, deciding for himself, decided as if for all.

These noble prospects drew him past pragmatic lures and empirical enchantments; they canceled narrowness and opened breadth. Kant glimpsed the inner bond between evil and self-destruction, between creativity and good. The occasional sternness of his tone and the constant rigor of his thought unluckily sustained generations of Prussian schoolmasters, who periodically yelped the platitude *Du sollst, denn du kannst*. However, the old philosopher had in fact quite another message, which was this: Men should not be treated as means merely, but rather as the ends for whom all human action is done.

It turned out, after all, that the ethics of command bore within itself, like hidden gold, the ethics of ends and means.

The legalism of the Kantian theory lies in its insistence upon a morality according to rule. The supreme imperative, described by Kant as "categorical," was self-validating like the law of contradiction. Consequently there was no need to rest the imperative's authority upon a divine being. In the Kantian system God, whose existence is beyond proving anyway, is left to preside over the ultimate adjustment of virtue to happiness.

Kantian ethics is thus in the main untheological. A legalistic ethics, however, need not be so. The sixth theory can be got from the fifth by simply adding the doctrine that the moral law requires proclamation by supernatural authority. On the principle of Ockham's razor (which, incidentally, has no foundation in logic) this view may be said to suffer from unnecessary assumptions. It has behind it, however, a venerable tradition extending back to the earliest codes in history; and, so far as the churches teach morality, this is the morality they teach.

(7) The seventh type of ethical theory is what we have already called the mountaineer's view. It holds that all value in the means (except the sole merit of efficiency) is derived from the surpassing value of one great end, one *summum bonum,* which lures the search and aspiration of mankind.

The chief problem in such an ethics is, of course, the definition of *summum bonum.* The various attempts at definition fall, generally speaking, into two categories. Some of them — a majority perhaps — take the phrase individualistically as signifying a superlative value which every man is to realize for himself. Aristotle, for example, seems to have defined *summum bonum* as happiness and happiness in its turn as the philosophically contemplative life. The late nineteenth and early twentieth centuries are littered with theories defining the highest good as some power or system of talents or harmony of structure in human personality. From Nietzsche's "superman" to Bradley's "realized self" the stress falls upon individual attainment, and I suppose that a similar ethic motivates the practice of psychiatry in the contemporary world.

The highest good can, however, be socially conceived, as it was by the Plato of the *Republic* and the *Laws.* The merit of this version is that it can give some moral import to self-sacrifice and, more generally, to the acceptance of risks on behalf of human progress. On purely individualistic theories the risks become quixotic and even perverse, since the more serious of them involve a frustration of the values which individualism sets itself to preach. You place Archbishop Cranmer in the flames, holding out his self-atoning hand, and you say that he suffered these torments because he "liked" doing so or earned thus a paradisical eternity or asserted thus the integrity of his character. Perhaps he may have done all these, but did he not also intend the bringing in of a new social order more conformable, as he thought, to the Kingdom of God?

The contest over definitions of *summum bonum* is in part a sophisticated and intellectualized reading of values actually in dispute in the given epoch. There is a danger, then, that the thinker may project as eternal what was only historical and temporary, and find himself holding sand instead of jewels when the struggle is spent. Accordingly, the definition of *summum bonum* needs to be made in terms of the historical process itself, so that the ideal indicates a quite practicable direction for change and can move events, like Aristotle's God, by attraction.

This kind of theory, moreover, corrects the other kinds. It supplies vision and achievement to the narrowness of immediate satisfactions and obliges both the hedonist and the intuitionist to search the future. It complements the one-sidedness of ethical legalism by attaching the rule to some great purpose which the rule is to serve.

The end-means concept thus appears to be at least an

ingredient in any persuasive ethical theory. One must confess, however, that *summum bonum* has been less adequately defined than any other philosophical term. One can, of course, always quarrel about definitions; it is the chief entertainment of intellectuals to do so. With most philosophical terms, however, such quarrels occur within limits, to exceed which would generally seem absurd. I doubt if this much can be said about definitions of the highest good. The difficulty suggests caution, but it will not prevent us from attempting, in the final chapter, a definition of our own.

The Mist and the Void

The last two types of ethical theory — (8) and (9) — are so different as to require a separate section. They are different in respect of the fact that they completely abandon the task which the other theories attempted, and they abandon it in the belief (or, as they would say, the knowledge) that it cannot be done.

(8) From the Greek thinkers through those of the nineteenth century the assumption is that ethical terminology is quite able to be defined and that the only subject of dispute is what the definitions shall be. In 1903, however, Professor Moore's *Principia Ethica* appeared, and altered the entire discussion. Moore didn't deny that there is a quality called "good," but he maintained that no definition can be given of the term. You may point to specific cases in which that quality is manifest, as you may point (it was his own illustration) to objects which are yellow. You cannot, however, give a *general* definition of "yellow" or of "good."

The reason for this, Moore thought, is that definition is a process of explaining complexities by resolution into simpler elements. Accordingly, when you reach the simplest terms of all, definition must stop. The simplest terms have no simpler terms by which they can be defined. If, as Moore maintained, "good" is one of these simplest terms, it will therefore be indefinable.

The theory seemed also to explain why previous attempts at definition had failed. The failure, indeed, appeared not merely in the existence of dispute. Moore was able to take the major theories one by one and show that each had committed what he called the "naturalistic fallacy" — the error, namely, of arguing from what is in fact approved to what ought to be approved. The commission of this fallacy was due, he thought, to the wish to define what is indefinable.

Now, it happens that "good" is used in a wide variety of meanings, not only by different people but by the same person on different occasions. The problem, then, is whether some one meaning can be established, such that we could say, "This is what people ought to mean by 'good,' no matter what they may happen to mean by it."

This problem is no mere exercise of mind. It is eminently practical. Suppose you belong to a group of people who are laying out a program of activity. Their combined activity might be directed to an indefinite number of ends, but among those ends they will have to choose just a few — or perhaps just one — as the object of their planning. They will have to decide which of the ends is "good" for them to pursue. Unless they all understand the term "good" in one and the same sense, how will they know that they agree upon their reasons for choosing? And unless a definition of "good" is possible, how will they know that they understand the term in the same sense? The effect of Moore's argument is to make rational social planning impossible.

If one is willing to say that Moore's view renders ethics a mist, one can go on to say that the mist covers a void. Ethical theory, having entered the cloud at the chasm's edge, did in fact plunge into the chasm itself. All that was required was to omit the notion that "good" signifies an intuitable quality, and the consequence was that "good" didn't mean anything except a casual, momentary, personal taste. This achievement, if collapse can be called an achievement, belongs to the logical positivists.

(9) We must try, now, to make as clear as possible what meaning it is that the positivists have caused to vanish. When people use the adjective "good" in its characteristically moral sense, they mean that such and such an act is one which ought to be done, that such and such an end is worthy to be desired. They do *not,* as a rule, mean merely that such and such an act *is* done or that such and such an end *is* desired.

Suppose we list a series of acts commonly regarded as bad: murder, theft, torture, drunkenness, incest. Every one of these acts has been at some time desired and committed by somebody. If "desired" and "worthy to be desired" were equivalent terms, and if both of them were equivalent to the term "good," then we should have to say that murder, theft, torture, drunkenness, and incest have been at some time or other good. Clearly we don't intend any such consequence. What we mean is that murder, though it may be desired, is not worthy to be desired, that a state of drunkenness, though somebody may want it, is not worthy to be sought.

Or again, suppose we list a series of acts commonly regarded as good: assistance to others, self-discipline, intellectual or artistic achievement. When we call these "good," we mean that they are worth desiring and worth doing, whether or not anybody actually desires to do them or does them. We assert their moral value regardless of what may in fact be wished or attempted or done. We mean that there is a rule, a standard of behavior which is binding upon everyone capable of moral action, and from the application of this rule we exempt only the moronic (who can plan nothing) and the powerless (who can effect nothing).

This *normative* meaning of "good" or of "bad" — the meaning conveyed by "ought," "should," "right," "wrong," "worthy," "unworthy" — is precisely the one which positivists deny to exist. According to them, if I state a moral judgment in the indicative mood, I am merely expressing a personal taste: "Candor is good" means "I like candor." And if I state the judgment in the imperative mood, I am merely expressing a wish about other people's behavior: "Be candid" means "I wish you would be candid" or "Would that you were candid." Sometimes a positivist will go so far as to construe such sentences in terms of social demands or prohibitions, that is to say, as mores. But I have yet to find a positivist who understands how mores are formed.

For a positivist, then, the cleavage between fact and value is complete. In the presence of disaster, in the midst indeed of horror and butchery, he can only sigh (and will let us only sigh), "I don't like this." And all the while the makers of disaster and horror and butchery answer, without interrupting their labors, "We like it very well."

Positivists have no morality: they cannot possess what they have destroyed. The lack is strange in them, because there never was a race of philosophers more devoted to hortatory prose. They are always upon you in every page, every line, nudging you, pushing you, sometimes encouraging but always threatening you, saying, "This you *must* believe!" But, my dear positivist, you just now "proved" that this only means you wish I would believe it. Suppose I don't wish to — and I don't — what will you do then? Call the police?

The fact is, however, that positivists don't like their moral nakedness, and when made aware that it is no Eden they stand naked in, they spend some time explaining that they don't quite mean what they say. In a recent and popularly written book, a leading advocate of the theory demonstrates in eleven pages that there is no ethics and in fifteen subsequent pages that one can be ethical anyway. He could not endure his own views. Having demolished morality, he wrote:

> Does that mean resignation? Does it mean that there are no moral directives, that everybody may do what he wants?
> I do not think so . . .[5]

But, of course, it does.

Defeat of positivism requires a demonstration of the ethical norm itself, such as we have promised to under-

---

[5] Hans Reichenbach, *The Rise of Scientific Philosophy,* University of California Press, 1951, p. 287.

take in the final chapter. Meanwhile, there are one or two more limited criticisms.

The first is that the supposed cleavage between fact and value is purest myth. The positivists know this perfectly well, or at least their behavior suggests that they do. For it wasn't mere perversity in them when they adopted the hortatory tone. They were writing down certain sentences which they asserted to be true, and to those sentences they invited our belief.

Nothing can be more natural. Every declarative sentence is, when asserted, a claim upon belief. It asserts a fact, to be sure, but it also invokes belief in the fact. That, you might say, is the human condition of sentences. And what is a sentence out of this condition? A sterile conglomeration of vocables.

Now, there is a gap between the stating of a fact and the invoking of belief in the fact. That gap is bridged by a tacit assumption to the effect that any statement which is true is worthy to be believed. Moreover, and in the same manner, there is a second assumption to the effect that any statement which is false is unworthy to be believed. Scientists have so long acted upon these assumptions that they have forgot they ever made them, and they particularly forget that they keep making them all the while.

These two assumptions are value judgments of precisely the kind which positivists deny: one's preference for true statements is not merely a matter of personal taste. If, however, the positivists are right, then these assumptions are meaningless. Then the gap between the true statement and one's belief in the true statement cannot be bridged. In other words, any connection between a true statement and our belief in it will be purely accidental and not due to the truth of the statement as such.

The situation is, however, worse than this. One can show that if the positivists are right, factual statements cannot even get themselves formulated. Suppose you undertake to describe accurately some portion of the world. You won't be able to do this very well at first glance. You will have to make observations and tests. How many observations and tests shall you make? Shall you average the results or draw inferences by sampling? In short, what method shall you use? Obviously, you have to make choices. But every choice is a value judgment. Consequently, without value judgments you will never reach the factual statements.

Thus it takes ethics to discover a fact and ethics to get the fact believed. The fantasy, which scientists spin, of concerning themselves with fact and not with value is as near to nonsense as rational beings usually attain. And the positivists, chanting this legend like Homeric bards, have justified Plato's worst fears about rhapsodists. I have heard that one of the most famous positivists, being asked what courses he wanted to teach during a certain semester, replied, "I want to teach the Truth." I applaud this remark, but I must insist that it is profoundly moral.

Thus nothing that actually goes on in the world supports the view that normative judgments are meaningless. What appears to be the case is, rather, that the procedure by which men conduct scientific inquiry is part of a larger method having to do with human behavior in general. The way we are to act in getting knowledge is a special case of the way we are to act in all our dealings with the world. Not only is science *not* separate from morality, but it is one of morality's most spectacular results.

The present prosperity of positivism is therefore not due to any evidence for the theory. It *is* due, however, to the society from which positivism springs and of which it is the faithful mirror-image. Positivism relies strongly upon the physical sciences; it is less knowledgeable about biology, and still less so about sociology. It has, as we see, no ethics at all.

Is this not a precise picture, a terrifyingly accurate picture, of our own commercial society? We know the vast physics necessary to control atomic energy; we know in lesser degree the biology needed to keep people in health. But we have not the social technology which would make these other sciences a blessing, and, our thoughts being more and more bent upon destruction, we have hardly any morality at all. Much physics, little sociology, and no morals: this is positivism, and this is what we are.

The condition, of course, is intolerable and cannot last. When the society vanishes which knew everything except what to do with its knowledge, positivism will vanish with it. Perhaps, indeed, it will vanish earlier, for it is incapable of conducting any sort of mass struggle against its various opponents. Thinkers of the twenty-first century, pondering gleams so tiny in our darkness, may well regard the positivists as children who, accustomed to candlelight, could not imagine that there is a sun.

Nevertheless there is one, and we can guess its radiance even before it dawns.

CHAPTER X THE HIGHER
HEIGHT

EARTH, THE WANDERER, had slipped through space many millions of years before some pregnant pool upon its surface yielded the first amphibian. Strange verdure covered the Devonian slime, where the leaves fell that were to turn to coal, and creatures crawled whose progeny were to walk. Then, for a long epoch, nature played with giant forms, and dinosaurs demonstrated the inconveniences of great bulk and small brains.

The centuries slid by to a sound of universal munching. Tyrannosaurus, with his six-inch teeth, gathered the fame and profits of a predatory life, while the gentle brontosaur, laying his head beneath the treetops, contemplated higher leaves. Then these astounding animals came to their sudden Mesozoic end. As the grass withered and the soil ran dry, the innocent and vengeful tribes alike learned the influence of economics upon history.

It is a million years or so since the saber-toothed tiger bequeathed his skeleton, prophetically, to certain pits of oil. A brief time later (as geological brevity goes) the exploratory ape dropped from his tree in Java to test the benefits of erect posture upon brains. Pithecanthropus was an ape trying to be a man, and he was wiser than some of his remote descendants, who, it would seem, are men trying to be apes.

The offspring of Pithecanthropus used their abler brains to discover the value of living in tribes. It was a discovery which not even the shrewdest individualist can annul. The half-human skeletons of the Chinese caves, lying amid the ashes of their borrowed and accumulated fires, attest a wholly human alertness to take what one cannot make. It remained for the first kindlers of fire and sowers of seed and architects of the wheel to add the *sapiens* to *homo* by showing how to make what one cannot take.

So there grew the latest animal, feebler than many, slower than most, bare beneath sunshine and before the frost, who nevertheless went everywhere, matched every power, did everything. Or, let us say, matched every power except his own, did everything except control himself. Despite all conquests of the physical world, man's social relations remained largely terra incognita until the nineteenth century. The inventors of society (as distinct from a condition of mere living together) did not beget a line of technologists comparable to those who followed the inventors of the wheel. The result is that man's ability to produce has been constricted by the social arrangements in which production occurs.

It is strange that civilization should have begun with the introduction of slavery, but that appears the probable fact. There came a time when human beings were able to produce more than each required for bare subsistence, and at that moment arose the possibility that part of a man's total product might go to support himself and the rest be appropriated by someone else. This condition happily solved the problem of what to do with prisoners of war. The victor set them at work for himself and lived upon their labor.

This was civilization as the Egyptians, Greeks, and Romans knew it, and they loved it with that special relish which ruling classes have for the structures which give them rule. Aristotle, who knew that slavery originated from war and was therefore outrageous, solaced himself with the reflection that some men are slaves "by nature." In this way, the Father of Logic became the father of all those preposterous theories which, in subsequent times, have defended inequality.

I suppose that even in our laggard age, when any view may flourish provided it is reactionary enough, there will be few to speak for chattel slavery. As a matter of fact, the ancients themselves maintained the institution by not discussing it. They fed it and enforced it, but they spoke of it with that prurient hesitation which hints a secret and astounding vice. They were certainly alarmed, in the first century A.D., at the rise of a new religion based upon the proposition that "the last shall be first." There were scuffles and persecutions, as when Pliny, under orders from Trajan, put to his harried Bithynians the question, "Are you now or were you ever a Christian?" [1] Then the recanters multiplied and the informers throve, and Pliny complained that in the welter of gossip he could find little to believe.

---

[1] "Interrogavi ipsos," wrote this elegant stylist, "an essent Christiani. Confitentes iterum ac tertio interrogavi supplicium minatus. Perseverantes duci iussi." (Pliny to Trajan, *Letters*, Bk. X, XCVI.)

The fears of the Roman imperium were not perhaps well founded. How was a mere procurator to know that the new religion, taking political facts as it found them, did not intend to effect the reversal of first and last with any undue celerity or indeed within the clouded confines of this world? It is very difficult for an official, when confronted with a mass movement, to note the refinements of its ideology. He knows that the movement moves and that it had better be stopped.

It was not Christians but barbarians who put an end to ancient slave society. When the barbarians had passed in their successive waves across the Empire, civilization wore the look of puddled ground after a great storm. Life resumed around the several pools which the tide had left. Central authority having vanished, local authority grew, and the beginnings appeared of a new system, the feudal, which was to last a thousand years.

The real conquerors had been those barbarian leaders who, acquiring estates in land, let out the farming of them in return for a certain number of days of free labor annually. From this relation came the complicated system of vassalage, in which an aristocracy of various ranks fed and fought, and a multitude of peasants tramped and toiled — a system so anarchic, despite its outward calm, that, while milkmaids dreamed in the fairy tales of marrying princes, the princes themselves, like Shakespeare's Henry V, could envy the simpler troubles of the serf,

> Who with a belly fill'd and vacant mind
> Gets him to rest, cramm'd with distressful bread;
> Never sees horrid night, the child of hell,
> But, like a lackey, from the dawn to set

> Sweats in the eye of Phoebus and all night
> Sleeps in Elysium.

One can pity the anguish of rulers who yearn, from time to time, to be someone else. But rulers make bad hedonists: they seldom allow the pangs of government to dissuade them from governing. For this purpose some more emphatic force is necessary, namely, a militant people, militantly organized.

By the fourteenth century a popular ground for social change was already developing. There were peasant revolts in both England and France, and Wycliffe made quite clear the future Protestant predilections of Englishmen. Feudalism fell, however, not by strength of the class which it directly exploited, but by strength of another class with which it merely interfered. The merchants rose upon two powers which feudalism itself had generated: discontent and technology. For gunpowder had nullified horse and armor, printing had shown that a book need not be chained in order to be read, and the compass and rudder guided men toward riches that only commerce might attain.

Sweet were those sweets of commerce, and gay the getting of them! Yet, like other sweets, they had a tendency to melt. Their color faded and their scent exhaled. No sooner had commerce sat down in the nineteenth century to enjoy its conquests over rank and space than some question arose over the digestibility of the meal.

> When the pie was opened,
> The birds began to sing.
> Wasn't that a dainty dish
> To set before a king?

In short, commerce sickened from its youth. Refusing a speedy end by starvation of profits, it had to accept a lingering death by surfeit of them. The scientific method, fatally introduced into physical knowledge, penetrated also into social; and for a hundred years now, despite all mystification, men have had sight of the truth that they must produce abundance collectively if they are to consume it individually. As Senator Robert A. Taft once observed, "Socialism in itself is not a bad thing." [2]

Ideals

The history of our race falls thus into three parts: man the animal, man the social animal, and man the socialist animal. During the first of these epochs men lived with little technology, during the second they established much mastery over physical nature, during the third (it is a conjecture partly demonstrated) they will establish similar mastery over their own social relations. There is a logic to this development, expressible in the single word "control."

If control of events is the basis of human freedom, in the sense that it makes possible an efficient satisfaction of needs, then freedom increases with control. History is thus seen to move, unevenly but steadily, in the direction of greater freedom; and it is difficult not to think of this development as *progress*, with all the overtones of moral value which that word conveys. This is because the movement shows evident signs of being a movement toward an ideal.

[2] "Education in the Congress," in *The Educational Record*, July, 1949, Vol. XXX, p. 351.

Without further argument, however, this impression of progress may be illusory. We can know that history is moving in a certain direction, and still be short of knowing that the direction is ideal. This purely philosophic doubt is much fortified by the presence in our society of powerful men who regard the socialist stage of history, not as progress, but as the dark and hideous end of everything good. It is fortified also by sight of the struggle and suffering which the transition seems to impose.

The question arises, therefore, whether there is an ideal, and, if so, what the ideal is. As we saw in the previous chapter, there are attempts by philosophers to dismiss the question as meaningless or not publicly answerable, but the necessity we are all in of making choices and (still more) the stress of social crisis raise the question oftener than it can be thrust aside. When oppressors assert that their ways are better and deserve to prevail, it is a pretty lame answer to tell them that the assertion is meaningless. What is required is a *demonstration* that their ways are not good.

Just as a matter of practical politics there is a tremendous loss of power in the abandonment of ideals, and it is inconceivable that masses could govern, or be governed, on the positivist view that the term "good" has no normative meaning. The normative part of the ideal is just what makes the ideal binding, and the effect is to make you accept the ideal as a rule of behavior without any direct threat of punishment. Surely, in this way rulers evoke from their peoples an immense amount of favorable action which, otherwise, they simply have not the physical power to coerce.

Yet, equally, if we imagine to ourselves a people truly self-governing, we shall find that a great deal of their behavior would have to be determined in much the same way. For in this case also they would have to accept an ideal as binding upon their actions, and submit their desires to it as representing the common good. You can escape the solitariness of being individual only by means of something which is universal; you can affirm your active membership in society only by accepting a rule, or set of rules, valid for all.

As a matter of fact, the same principle will appear even if we limit the scope of ethics to the individual person taken by himself. Every such person, I imagine, tries to effect some sort of unified development of his own life. He thinks of his schooling as related to a job or career, the job or career as related to marriage and the hope of children, and these in their turn as related to a secure old age in which he may enjoy the gains of labor and fecundity. The stages of life thus related constitute a system. There is a principle running through them and a rule of action following from them, and to that rule the man must submit his desires just as surely as if he had been considering society's fate rather than his own. Even an opportunist, though he has (by definition) no other principles, has at any rate the principle of opportunism. Even the obsequious slave of social custom credits the value of his slavery.

The continuity of social life and the continuity of personal life are both examples of permanence embracing change. The human values which invest this continuity are not exhausted by any pleasurable moment nor by any sum of pleasurable moments. This is the fatal defect of all hedonism, of every ethics based upon brief adaptation or temporary command. The *process* escapes them. They are anthills, and they may be swarming, but they are not the world.

Ideals, therefore, must be large enough to be encompassing. They lie beyond the remotest future; and, if we can suppose that their plurality is composed into a unity, the one supreme ideal has probably the nature of a limit which can be approached but not attained.

The fact that no value can be exhausted within one moment and that nearer goals turn out upon attainment to be steppingstones offers a certain lure for transcendental moralizing. Ideals, it may appear, are always relevant to immediate experience, while yet eluding it. Value lies along the side of fact, lies also above it and around it, making it bland and sometimes glorious. The Hegelians, as we have seen, suggest that man already lives in heaven, shut out from its radiance, however, by simple finitude.

Yet, if we are to have two worlds in this way, with science addressing itself to one and ethics to the other, there will be (as there have been) endless disputes about the rival jurisdictions. Fact will try to devour value in the positivist manner; value will try to devour fact, in the Hegelian. Still worse, we shall be burdened with the dreadful metaphysics of explaining an order of existence to which sense experience has no access. Hegel attempted this by logic, and Kant by pragmatic conjecture. The common failure of these two great minds leaves no invitation for further effort.

In saying this, we are accepting the positivist critique of transcendental ethics, and are agreeing that there isn't any order of existence beyond space and time. But we don't think it necessary to pour the baby out with the bath water. On the contrary, we hold that, in the world of space and time, values can be found having their expected normative character, binding upon choice and worthy to be achieved.

The Sociology of Moral Skepticism

The belief that this task cannot be done is still dominant in Western philosophy, though it does seem weakening a little under stress. To judge by their recent books, the positivists appear to be rediscovering ethics, just as the pragmatists, confronted by the Lysenko controversy, suddenly rediscovered "eternal truths." These weakenings are no doubt the first obvious effects of new political pressure, and there will be more of them. Meanwhile, we address ourselves to the dominant view.

Moral skepticism — the belief that no ideal or rule of action can be proved valid for everybody — rests upon one (or possibly all) of the following grounds:

(1) The rules which we call ethical, it is said, are in fact merely the reigning habits of a particular society. The feeling of obligation which they arouse is nothing moral at all, but reflects the threat of public disapproval or coercion. The most that ethical theory can do is to tabulate these customs as you would tabulate votes — recording, let us say, one hundred million for the Ten Commandments, fifty million for pleasure, and one million for the categorical imperative. "Good" therefore means "approved by public opinion," and people like

Socrates who flout public opinion are seen to behave in a very quixotic way.

(2) The adjective "good" cannot be defined, although the substantive phrase "The Good" can be defined. This was the doctrine laid down by Professor Moore in 1903. It is regarded by philosophers as one of the classic formulations, but I suppose it has had no popular currency at all. It asserts (a) that you can say, "For the purposes of the following ethical theory the term 'The Good' will mean such-and-such"; but (b) that you can show no necessary relation between this definition and what people ought to mean when they describe a person or a state or an action by the adjective "good." I have to confess that this doctrine seems to me almost entirely verbal. It is not altogether clear to me, nor is it clear to me that it is clear to Moore. Nevertheless, discussion of it is *de rigueur*.

(3) That the meaning of moral statements is either factual or expressive of personal wishes and commands. Thus "You ought not to stay here" means "I don't like your staying here" or "I wish you wouldn't stay here" or "Go away." I suppose that, correspondingly, "Kite-flying is good" means "I like kite-flying" or "Go fly a kite." The doctrine has been somewhat popular since Mr. Stuart Chase discovered its existence in 1937. Since that time, quite a few intellectuals, when confronted with a clash between moral imperatives and safety, have found it useful to reflect that such imperatives merely state somebody's private wish.

If you review these doctrines for a moment, you will observe that the first and third resolve value completely into fact (into sociology and psychology respectively), while the second believes value to exist but cannot describe it. Two of them can talk, but have nothing to say; the other has something to say, but is tongue-tied.

Now, there is no objection in principle — none at least that I would have — to a relation between sociology and ethics. But the sociology invoked by these theories bears only a distant resemblance to the sociology that is. The class structure of society and the conflicts of classes are all absent from it, and economics seldom rears its formidable head. True to their nominalist logic, the theories regard society as a congeries of individual persons, whose behavior is not much influenced by the social relations they stand in. Consequently, the theories are poorly fitted to understand history as a complex and uneven movement toward an ideal. Indeed they do not even guess that their own content and acceptance is prompted by social causes.

In this last fact there is some irony, but there is also much explanation. For when you consider the matter, you realize at once that it must be impossible for competitive societies to generate agreement in ethical theory. Theoretically they have a set of rules to compete by, but in practice competition effects a constant breaking of the rules. Moreover, "competition" is a euphemism for what actually goes on: in fact there is a struggle, always involving violence, by which ruling classes try to maintain their rule, and the members of ruling classes try to stay within their class. From this arise two sorts of ethics: one which justifies the power of rulers and seeks to make their purposes seem socially enlightened, and one which expresses the hopes of submerged multitudes. Unanimity is impossible.

In this welter of opinion about the meaning of "good," the very quantity of conflicting definitions may well make it appear that "good" is not publicly definable. But by looking at the class structure of society we can see exactly what section of the community it is in which such a belief is most natural. Further, when we have seen the clear source of the belief, we shall recognize that there is even less reason for holding it.

Suppose that, in the Marxist manner, we regard Western society as basically divided into capitalists and workers, with colonial peoples suffering the intensest exploitation of all. Then I think it will appear that each of these groups understands the adjective "good" in a sense appropriate to its class position. That is to say, capitalists need profits, workers need steady ample wages, and colonials need at least subsistence upon their own land. Any professions to the contrary will speedily be confuted by the actual behavior of the groups. I think it is fair to say that capitalists understand "good" in a sense common to capitalists, workers in a sense common to workers, and colonials in a sense common to colonials.

Who, then, are the people for whom "good" has no definable meaning and ideals have no validity? They are the rootless people who have no unseverable ties to any of the great class divisions. They float upon the social tide like ambergris, cast out by Leviathan but of great perfume. Chief among them are the intellectuals, the migratory workers of the mind, who sell their services to the capitalists but who are rarely admitted into the class itself. Equally they are not, by habit of life, proletarians or colonials, though their origins may lie

THE HIGHER HEIGHT

in those groups. Their salaries and commissions bind them to the dominant class, from which they take whatever color they have.

An intellectual's one duty is to have convictions. If he has no other conviction, he must at least have the conviction that one ought not to have convictions. Whatever the convictions are, they ostensibly result from study and reflection, and, as a matter of fact, do so in part result. They are, however, strongly modified by the need of maintaining income, of having access to the channels of communication, of possessing such fame as tolerable talent may win. Toward the less privileged classes intellectuals feel free and even irresponsible: the struggles of workers and colonials leave them with a belief in art for art's sake.

A militant anarchy exists among them, reflecting their competition for favor. They divide into schools, it is true; they establish trends, and they even move in movements. But they are solid atoms of belief: like Epicurus' particles, they bounce upon collision and never merge. They have various functions, random and salaried, but they have no single social function which they can feel to be historically ordained.

What will "good" mean to them? Well, in the innocent years of their training, it will signify the passing gleam of this or that insight, the lucky refinement of this or that taste, the prospective career of a rolling atom more compact of thought and more lustrous of surface than any that have rolled before. Once the career begins, the atom's chief care is to roll without shock. It gathers knowledge, surely, but the most important of its knowings is the knowing what theories are safe. Consequently, it rolls in various directions, avoiding with an acquired but almost preternatural skill the monster atoms which would destroy it. It has no *tendency,* no historical mission, and its life is the sum of its escapes.

In this tumult of variety and evasion intellectuals soon grow impressed with anarchy of opinion: there were differences of view throughout the past, there are differences in the present, there are differences indeed within one's private mind. Thus the wavering or (as they now love to call it) the "ambivalence" of intellectuals follows from their very conditions of life. No doubt they share the common human necessities, but even these they sometimes discover only with psychiatric help. It is people like these, the occasionally useful by-products of history, who proclaim the indefinability of "good" and the invalidity of ideals.

The philosophical reasons for moral skepticism are no doubt subtle, and for twenty-five years they have been subtly advanced. But, for my part, I don't believe that the philosophical reasons are the real reasons why the theory is held. Rather, they are reasons you would accept if, and only if, your career in society were such as to dispose you toward them.

At any rate, these are the social roots of moral skepticism. Before the graceless infirmity of the doctrine itself even its blithest propagators may wince. Its role in history is ending now, and the time has come to remove the theoretical supports. This we can do by producing a definition of "good" and a demonstration of the ideal.

The Highest Good

In prosecuting the task thus set us, we shall begin by seeing whether we can remove the grounds on which Professor Moore decided that the adjective "good" is indefinable. That done, as I hope it will be, we can then go on to a definition of "good" and a formulation of the ideal.

(1) Moore's notion is that you can define a term provided it is the name of a complex entity; the definition then consists in indicating the parts and their arrangement. "Good" he holds to be, as adjective, the name of a simple quality. It has, accordingly, no parts which can be enumerated; and thus no definition is possible. The same thing holds, Moore says, for the sense-quality we call "yellow." You have to intuit the quality directly in order to know it, and you can frame no definition of it which would convey its nature to a blind man.

Now, I think that, so far from showing that there is no public definition of "good," this analogy shows that there is at least a part of one. If you want to convey to someone else your understanding of the word "yellow," you will point to something exhibiting the quality — a bunch of daffodils, say, or a heap of sulphur. But when you do this, you must also say that you are pointing out a *color*, that is, a certain *type* of quality which things have. If you don't indicate this particular type of quality, your companion won't know whether to look for color or shape or size or something still else. For all he can tell, "yellow" may mean "pyramidal" or "flowerlike."

In naming a type (and you must name one) you have got half of what all logic books since the time of Aristotle declare to be formal definition. You have got the *genus* or large inclusive class in which will be found the thing you propose to define. All that is lacking is the *differentia,* which will distinguish the thing you are interested in from the other members of the class.

If, now, we proceed with Moore's analogy, we find that unless you say what type of quality "good" is, your pointing to an example of it will be mere obfuscation. But if you can, and do, name the type, what is to prevent your stating the *differentia?* If you state that too, you have a full public definition. It certainly seems as though one could try.

Suppose we begin where Moore begins and say that the term "good" refers to a quality found in people, things, acts, and states. Suppose, further, that we repeat our earlier description of the universe and say that people, things, acts, and states are all processes which modify one another. Then we ought to be able to reach the quality "good," as we would reach any other quality, by examination of the context in which it characteristically appears. We would get to know it in the same sort of way we get to know a man by his relations with his fellows.

Since the universe is spread out in time, a process moving into the future, some of the qualities a thing has are due to what it will become. This is most obviously the case with those things and acts which owe their nature to human purposes. Here the future influences the present as a blueprint influences a designed machine: it is prophecy teaching fulfillment. Every stage of the construction is what it is because of the final effect which is to be.

At the same time, there plays throughout the process a certain amount of causation from the past, joining with the future at just the moment where we now are, as if the past had thrust us forward and the future laid a hand behind us to draw us on. In this way the continuity of events survives, however new and strange the stages of them.

In human life this continuity is effected by certain needs, which are past and present facts envisioning and demanding a specific kind of future. If I want to become a physician, my being admitted to medical school has a very different quality from what it would have otherwise. That new quality is the quality "useful to me"; and if I wanted to define it, I should have to do so in terms of the whole context.

The *whole* context, however, and nothing less. Applying this rule to the definition of "good," we find ourselves obliged to consider all mankind in its full historical relationships. The definition of the term merges with the statement of the ideal, and this is the second part of our task.

(2) Let us say, as a convenient scheme, that the context we are here studying has a vertical and a horizontal axis. Let the horizontal axis represent mankind taken as a whole in any one set of historical relations. Then the vertical axis can represent the development of mankind through all those sets. My view is that the two axes, taken together, will enable us to reach a formulation of the ideal.

First, the horizontal axis. Here the fundamental lesson is that history is made by masses. No lord or slaveowner, no magnate or imperialist, no bearer of exalted titles — none, in fact, of that stupefying company who have thought from time to time that history would answer their slightest summons — has done more than contribute to the process. What happens in history is the effect of multitudinous wills, which, acting upon one another, modify all intended consequences. From this follows the curious aspect that history shows, of seeming impersonal though made by persons. An employer, one might say, "wills" profits; his employees "will" high wages; his competitors "will" an increasing control of the market. The result precisely corresponds to none of these intentions taken separately, but shows very plainly how the intentions have modified one another.

This "impersonal" nature, we may observe in passing, is the reason why history can be a science with the predictive power which sciences have. If historical events represented nothing but the random caprices of individual men, the success or failure of Cleopatra's policies might be explainable by the length of her nose. But the pressure of other interests upon policy is always great enough to remove anything adventitious, and so it becomes a first principle of statecraft to formulate policy without caprice. This or that foreign policy (for example) may seem truculent, but it is never the result of mere intestinal malaise.

History has been made by masses, but until quite recent times the masses were separate groups making, so to say, separate histories. The horizontal connections were interrupted in the older epochs of narrow parochial living: Europeans of the tenth century had no relations with their contemporary Aztecs. I imagine that the actual disconnection of mankind did much to invest the ethics of brotherhood with an air of unreality. One could not, except imaginatively, describe the unity of mankind; and one's notion of social welfare shrank to the limits of a city-state, a Mediterranean empire, a "universal" (i.e., merely European) church.

However, a quite new era came when, during the nineteenth century, commerce unified the world. Since then, our horizontal connections have stretched around the globe; the straight lines of *our* sociology are all curves. Nothing can happen anywhere, especially in an economic way, which has not some lively influence upon us all. The unity of mankind is almost a present fact and very palpably a future one. What the nation-state hid from Hegel, what competitive society hides from the nominalists, is just this oneness, in which lies the definition of "good" and the validity of ideals.

The interaction of wills, as we were describing it, takes place within an inherited system of relations, which can be called successful to the extent that it enables men to satisfy their needs. Whenever a system fails to do so, the cumulative strivings of men proceed to alter it: in this way, for example, feudal society got turned into capitalism. The new system, toward which the altering moves, is partly determined by the vanishing system and by the social forces which it generated. This gives us the vertical axis, the dynamic of history.

It appears that in all epochs men have needed food, shelter, play, companionship, and love. These are the rudiments of security, the assurance that in this kind of animal at any rate there is something more than a dull nosing over the surface of earth. Needs, in their turn, develop wishes, and the wishes plans. Into both these stages, of which the one is conscious and the other fully deliberative, distortion sometimes enters. All needs are legitimate and deserve satisfaction, but some of them get twisted into mistaken or even malignant wishes, and the wishes into noxious plans. Nevertheless, the conscious wish and the deliberated plan are the source of effective control over environment. Without them we would have been the feeblest species in the field.

The course of history seems also to show that men generally refuse to surrender their social gains, their increased efficiency in satisfying needs. These gains, indeed, they are so well aware of that they experiment very cautiously with the future, for fear of losing them. It is a powerful persuasion toward the *status quo*, and I think that right now most Americans prefer a bird in the capitalist hand to two in the socialist bush. You change your mind only when there are no longer any birds.

Besides all this, men show a concern for the perpetuation of their race. It is in them a sort of inescapable altruism. The concern is genuine, even if momentarily obscured by the atomic bomb and Operation Killer and headlines like the following: STOCKS TUMBLE SHARPLY ON LATE PEACE RUMORS.[3] The forward generations, however, contemplating this madness retrospectively as we contemplate the madness of Caligula, will show by their existence and progress that we intended them to *be*.

[3] *The Philadelphia Inquirer,* March 15, 1951, p. 36.

These arguments bring us within sight of the ideal, but there is one thing more to do before we formulate it. Let us take the ethical theories mentioned in the previous chapter and express them, illustratively, in economic terms, reading their moral values for the moment as economic values. Thus expressed, they will be found to divide into a consumer's point of view and a producer's point of view.

The consumer's point of view is most obvious in the pleasure theories, which identify value with enjoyment. Stoicism is the attitude of consumers in extreme scarcity. Legalistic theories deal with the problem of fair distribution, and cultural relativism shows contentment with distribution as it is (at least in foreign societies). The intuitionist view, lastly, leaves it to the private person to improvise his own rules of possession.

At the same time, the legalistic theories, because they emphasize laws, codes, or rules, are also concerned with procedure. They are concerned, that is to say, with methods as much as with results. Kant, indeed, in his stricter moments professed to be concerned with methods alone. This is a producer's point of view. It degenerates, when severed from consequences, in the same way that production degenerates when severed from social use. In its more intellectual forms it becomes casuistry, and in Dewey's instrumentalism it shows the American taste for gadgets.

But human beings are producers and consumers; they are inseparably both. The values they seek, admire, and accept are organized precisely according to this relation. Men praise the masterpiece *and* the genius of its source. The creator is valued in his creation, and the creation in him. You cannot break this relation without shattering it irrecoverably.

For this reason it seems necessary to project the entire fabric of society — that is to say, mankind in all its basic relations — into the ideal. The ideal is therefore a mass ideal, displaying not abundance alone but the flowering of all skills which make abundance possible, an harmonious adaptation of production to use.

Drawing, now, all these arguments together, we are ready at last to state the ideal. Since an account of perfection must necessarily be couched in superlatives, we can dismiss the cautiousness of mere positive degree, and define the ideal thus: *The moral ideal is that organization of mankind in which the satisfaction of human needs occurs with complete efficiency.*

Apologia

Having now stated the ideal, I must say that I do not suppose it will please everybody (though why it shouldn't I really don't know). Nor do I suppose that the argument is definitive. There are gaps in it, and there may also be (though innocently) traps. Nevertheless it has a cumulative force which I hope may prove difficult to resist.

There is, of course, one type of person who is granite to all such pleas. He is consciously antisocial or contentedly calm, and, try as you may, you will find no arguments to persuade *him*. Even if you show that a denial of your ideal renders his own actions inexplicable, it makes no difference: he doesn't want his actions explicated anyway. After this same manner Professor Tawney wrote, very wittily:

> There are many, of course, who desire no alteration, and who, when it is attempted, will oppose it. They have found the existing economic order profitable in the past. They desire only such changes as will insure that it is equally profitable in the future. *Quand le Roi avait bu, la Pologne était ivre.* They are genuinely unable to understand why their countrymen cannot bask happily by the fire which warms themselves, and ask, like the French farmer-general: "When everything goes so happily, why trouble to change it?" Such persons are to be pitied, for they lack the social quality which is proper to man. But they do not need argument; for Heaven has denied them one of the faculties required to apprehend it.[4]

Well, then, we won't address ourselves to this sort of people, but at the same time we must beware of including all objectors in this category. There surely are some people who would doubt our formulation of the ideal, not because they are self-interested, but because they are unconvinced. They deserve further answers, if they wish them; and no doubt our theory would gain by the discussion.

Philosophers, by long habit, place their faith in abstract demonstration; and their first impulse, in defending an ideal, would be to argue that the denial of it leads to self-contradiction. This procedure is by no means unimpressive: if the denial of a certain statement does involve logical inconsistency, then the denial cannot be rationally maintained and the denied statement must be accepted. It would be pleasant to think that this condition holds for our ideal.

[4] R. H. Tawney, *The Acquisitive Society*, Harcourt, Brace, 1920, p. 3.

I am afraid, however, that it does not. An objector wouldn't need to hold that our ideal has no validity at all; he would only need to hold that it has not *complete* validity. Then he would be saying something like this: "It is good, but it is not of the *highest* importance, that there should exist an organization of mankind in which human needs are satisfied with complete efficiency." There are in fact a number of such theories, and among them is Calvinism, for example. A Calvinist wouldn't deny the desirability of social perfection, but he would say that there is something still more important than this, namely, the glorification of God.

Now, Calvinism is a particularly good case of systematic theory. It has been worked out in detail, and, so far as I know, there are no inconsistencies in it. You may say that some of its assertions are false, but I doubt you will be able to show that any of them contradict any others. This being true, it must be the case that a denial of our ideal does not necessarily produce illogical results. And even if we could claim that it did, we would still not silence all possible objectors, who might then take their stand on mystical grounds alone.

What we can claim, I think, is that to deny our ideal is to make the separation of theory from practice complete and final. In other words, an objector, maintaining in theory that the efficient satisfaction of human needs is not of the highest importance, would nevertheless be acting all the time as if he thought it was. We would find him satisfying *his* needs as efficiently as he could. He wouldn't miss his dinner "for the world," and he would spend as little effort in getting it as the circumstances permitted. His practice would deny what his theory asserts.

He might of course maintain, but as a respectable citizen he probably wouldn't, that the ideal, though it holds for him, does not hold for everybody and so is not universal. Then he would be saying that there are at any rate some people whose needs don't require efficient satisfaction. And this, I suppose, is the actual position of men who make their wealth out of other people's labor, sometimes very ruthlessly indeed.

But this position, though not inconsistent with a certain kind of practice, does contravene the *social* practice which has evidently been developing for centuries. It contravenes, that is to say, the entire historical development in which men, experimenting with different skills, have also been experimenting with various forms of social organization, seeking (as it were) the form in which technology and abundance are most happily combined. Here the separation of theory from practice takes the form of a collision between the practical theory of a minority and the evolving practice of mankind as a whole. It interrupts, and a little postpones, progress toward the stage in which social planning makes men masters of their destiny.

The separation of theory and practice is a primal fault, requiring, for amendment, some effort on the part of both. While it lasts, all plans must be somewhat futile and all actions somewhat self-defeating. But amendment also requires an harmonious society, in which the needs of members no longer generate mere personal excuses but are grasped and gratified within a theory of the whole. If, by this doctrine, our ideal stands justified, I am more than content. For then there is healing in these pages, philosophy displays its vindication, and the promise to Mrs. Nixon is redeemed.

I suppose that pain and some few terrors lie between us and our goal. A strife of epochs, such as now seems our lot, has the proportions of discord within the firmament. Some stars will fall — the baleful ones, the sickly constellations — trailing their flames across the finite universe until they come to its edge and vanish in the void beyond.

But the earth, as I imagine, will remain, turning steadfastly upon its axis. The shadow of its turning will be time for love and sleep, and its turning from the shadow will be time for work and joy.

INDEX

INDEX

ABSOLUTE, THE, 52–65, 70, 82, 89–91, 103–104
"Accommodation," 26–32, 79
Acheson, Dean, 114
Agnosticism, 30–31, 48
Aiken, Conrad, 28
Alexander the Great, 21
Anarchism, philosophical, 89–90
Anaxagoras, 20
Anaximander, 130
Anaximenes, 54
Aristippus, 216
Aristotle, 17, 21, 66, 122, 129, 192, 193, 226, 227, 238, 252
Arnold, Thomas, 47
Atheism, 30–31
Augustine, 17
Authoritarianism, 145, 155–163, 169, 170, 171–172, 173–175

BENTHAM, JEREMY, 56, 217–218
Berkeley, George, 104, 110–112, 186–187
Blake, William, 167
Blakeley, Robert, 47
Bosanquet, Bernard, 56, 59, 61, 71–74, 75–80, 136
Bosanquet, Helen Dendy, 75, 76
Boyle, Robert, 14
Bradley, Francis Herbert, 45, 57, 61, 70, 75, 90–91, 136, 139, 226
Brockmeyer, Henry C., 49
Brooks, Van Wyck, 28
Butler, Joseph, 149, 219

CAPITALISM, 12–13, 29–31, 52, 55, 69–74, 76, 120–121, 160, 192, 218, 235, 240–241, 248, 252
Caligula, 256
Calvinism, 260
Carlyle, Thomas, 37, 48
Carroll, Lewis, 152
Change, 35–42, 70, 75, 121–140
Chaplin, Ralph, 120–121
Chase, Stuart, 28, 246
Christianity, 238–239
Cleopatra, 254
Cleveland, Grover, 87
Coleridge, Samuel Taylor, 48
Colonialism, 25, 65, 77, 120–121, 248
Copernicus, 191
Cousin, Victor, 46
Cranmer, Thomas, 227

DARWIN, CHARLES, 47, 155, 221–222
Descartes, René, 14, 24, 32
Dewey, John, 28, 83, 257
Dialectics, 124–140
Diederichs, D. N., 58
Dionysius the Areopagite, 163
Dreyfus, Alfred, 87

ECONOMICS, 12–13, 222
Einstein, Albert, 155
Emerson, Ralph Waldo, 49
Empiricism, 46, 97, 193–200
Engels, Friedrich, 42, 48, 139–140
Epicurus, 216–217

INDEX

Epistemology, 8, 35–36, 96–101, 136–137, 143–206
Error, social origins of, 22–32, 182–183, 204–206
Ethics, 8, 36, 38, 60–65, 155, 205–206, 210–262
Evolution, 237–241

FARRELL, JAMES T., 28
Feudalism, 29–30, 113, 160, 192, 239–240

GALILEO, 191
Gladstone, William Ewart, 47, 75
Good, definitions of, 213–235. See also *Summum bonum*
Guedalla, Philip, 46

HAMILTON, SIR WILLIAM, 48
Harris, William Torrey, 49–50
Hazlitt, Henry, 28
Hedonism, 216–219, 244, 257
Hegel, G. W. F., 41, 45, 50, 61, 71, 90, 95, 104, 108, 110, 111, 126, 139, 244, 255
Hegelians, 41, 45–80, 84, 97, 102–104, 136, 138
Heraclitus, 19, 128, 131–133, 134
Herodotus, 221
Hicks, Granville, 28
History, 237–241, 253–256
Hobbes, Thomas, 14, 216
Homer, 132
Horace, 10–11, 19, 216
Humanism, 27–32, 37
Hume, David, 4, 48, 88, 111, 139, 196
Huxley, Julian, 28

IDEALISM, 41, 67, 68–69, 109, 110–116, 135–136, 241–245, 251
Intellectuals, 25, 135–136, 248–250
Intuitionism, moral, 219–220, 257

JAMES, WILLIAM, 41, 50, 82–101, 106, 111, 115, 122

KANT, IMMANUEL, 9, 17, 88, 139, 223–226, 244, 257
Kepler, Johannes, 191
Kierkegaard, Sören, 172
Kipling, Rudyard, 87
Klem, William, 186–187
Knowledge, theory of. See Epistemology
Koegel, Otto E., 68–69

LEGALISM, ETHICAL, 223–226, 257
Leibniz, G. W., 14, 191
Lerner, Max, 28
Lewes, George Henry, 48
Lippmann, Walter, 28
Locke, John, 14, 17, 23, 100, 139, 193, 204, 221
Loyalty, 64–65

MACHIAVELLI, NICCOLÒ, 216
MacLeish, Archibald, 28
Mann, Thomas, 28
Martineau, James, 215
Marx, Karl, 13, 42, 71, 90, 108–109, 139, 205
Materialism, 109, 111
Methodology, 137, 189–206
Milton, John, 140, 177
Moore, George E., 219, 228–230, 246, 251–252
Moran, Charles, 186–187
Mysticism, 117–118, 145, 163–169, 170, 172–175

NEWTON, SIR ISAAC, 14, 47, 155, 191
Nietzsche, Friedrich, 226
Nixon, Mrs. Isaiah, 33–38, 59, 62, 91–92, 102, 170–173, 177, 211, 262

O'CASEY, SEAN, 148
Ontology, 8, 45–140

PALMER, GEORGE HERBERT, 50, 51
Parmenides, 4, 52, 57, 122
Pascal, Blaise, 99
Permanence, 37–42, 45–80, 130

INDEX 267

Philosophy, nature of, 7–18, 20–21
Pius XII, 107–109
Plato, 10, 17, 21, 40, 66, 88, 126, 226, 234
Pliny, 238
Politics, 12–16, 66–74, 161–163, 242–243
Positivism, 21, 105–106, 109, 139, 230–235, 242, 244–245, 246
Pragmatism, 21, 68–69, 82–83, 102–103, 105–106, 109, 136–137, 138, 199–200. *See also* James, William
Protagoras, 20–21
Pyrrho, 146, 150

REALISM, 82, 136, 138
Reichenbach, Hans, 232
Relations, internal and external, 125–128
Relativism, 143–153, 169, 170–171, 173–175
Royce, Josiah, 50–51, 53, 55, 61, 62–65, 79, 83–84
Russell, Bertrand, 28, 83, 139

SAINT CATHERINE OF SIENA, 166
Saint Francis of Assisi, 165
Santayana, George, 28, 83, 93, 136
Schopenhauer, Arthur, 39
Shakespeare, William, 123, 176, 212–213, 239
Smith, Warren Allen, 27–30
Socialism, 29–30, 77, 90, 192, 241, 242

Society, 22–27, 191–193, 238–244, 245–258
Socrates, 17, 21
Sorel, Georges, 78
Spencer, Herbert, 48, 77, 85, 218
Spinoza, Baruch, 10, 14, 17, 88, 118, 195
State, the, 70–74, 121
Stephen, Leslie, 64
Stirling, Hutchison, 48
Stoicism, 210, 259
Substance, concept of, 128–130, 134–135, 137–138
Summum bonum, 226–228, 251–258
Supernatural, the, 24, 103–105, 107, 116–121, 164, 183, 226

TAFT, ROBERT A., 241
Tawney, R. H., 259
Tennyson, Alfred, Lord, 47, 166–167
Thales, 20, 128–129, 134
Theory and practice, 193–200, 260–262
Trajan, 238
Truman, Harry S., 67, 176
Truth, nature of, 178–188

UTILITARIANISM, 56, 217–219

WHITEHEAD, ALFRED NORTH, 42, 137–139, 196–197
Wycliffe, John, 240